Exercises in Helping Skills: A Manual to Accompany

The Skilled Helper
A Problem-Management and Opportunity-Development Approach to Helping

NINTH EDITION

Gerard Egan

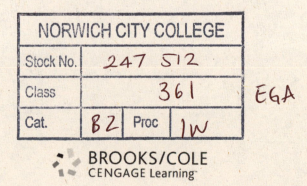

BROOKS/COLE
CENGAGE Learning

Australia • Brazil • Japan • Korea • Mexico • Singapore • Spain • United Kingdom • United States

BROOKS/COLE
CENGAGE Learning

For product information and technology assistance, contact us at **Cengage Learning Customer & Sales Support, 1-800-354-9706**

For permission to use material from this text or product, submit all requests online at **www.cengage.com/permissions**
Further permissions questions can be emailed to **permissionrequest@cengage.com**

ISBN-13: 978-0-495-80632-5
ISBN-10: 0-495-80632-3

Brooks/Cole
10 Davis Drive
Belmont, CA 94002-3098
USA

Cengage Learning is a leading provider of customized learning solutions with office locations around the globe, including Singapore, the United Kingdom, Australia, Mexico, Brazil, and Japan. Locate your local office at: **www.cengage.com/global**

Cengage Learning products are represented in Canada by Nelson Education, Ltd.

To learn more about Brooks/Cole, visit **www.cengage.com/brookscole**

Purchase any of our products at your local college store or at our preferred online store **www.ichapters.com**

Printed in the United States of America
1 2 3 4 5 6 7 13 12 11 10 09

To Hany, Laura, and Rich. Great colleagues. Invaluable friends.

Thanks, Gerry

CONTENTS

INTRODUCTION
Getting the Most Out of These Exercises

These exercises are meant to help you translate concepts in *The Skilled Helper* into skills you can use. If you already possess some of the skills, these exercises will help you improve them and integrate them into the helping model.

When it comes to learning skills, there is no real substitute for practice. You can practice in three ways—first, by doing these exercises on your own; second, by sharing your responses with your fellow learners; and third, by applying what you learn to your everyday personal and social interactions outside of the classroom. In the following pages you will find a blend of in-class exercises, as well as exercises that focus on your daily life outside of the classroom. Classroom activities are important because they provide opportunities to practice, in addition to giving and receiving feedback.

A TRAINING PROGRAM FOR HELPERS

As the title *The Skilled Helper* suggests, helpers-to-be need a helping model as well as a range of techniques and skills to make that model work. These skills include basic and advanced communication, the ability to establish working relationships with clients, and the ability to help clients identify and explore problem situations and unused opportunities, set problem-managing and opportunity-developing goals, develop action plans to accomplish these goals, implement plans, and engage in an ongoing evaluation of this entire process. These skills are not just pieces of helping technology, but rather creative and humane ways of helpers' total investment in their clients' efforts to better their lives. The only way to acquire these skills is by learning them experientially, practicing them, and using them until they become second nature.

You are about to embark on a training program in counseling. The following are standard steps you will encounter in such a program:

1. **Cognitive Understanding**. Develop a cognitive understanding of a particular helping method or the skill of delivering it. You can do this by reading the text and listening to lectures.
2. **Clarification**. Clarify what you have read or heard. This can be done through interactions with classmates and instructor-led questioning and discussion.

The desired outcome of Steps 1 and 2 is *cognitive clarity*.

3. **Modeling**. Watch experienced instructors model the skill or method in question. This can be done "live" or through films and videotapes.
4. **Written Exercises**. Do the exercises in this manual that are related to the skill or method you are learning. The purpose of this is to demonstrate to yourself that you understand the helping method or skill well enough to begin practicing it. The exercises in this manual are a way of practicing the skills and methods "in private" before practicing them with your fellow trainees.

The desired outcome of Steps 3 and 4 is *behavioral clarity*.

5. **Practice**. Move into smaller groups with your fellow trainees to practice the skill or method in question.

6. **Feedback**. During these practice sessions, evaluate your own performance and get feedback from a trainer, as well as from your fellow trainees. This feedback will confirm what you are doing right and correct what you are doing wrong. The use of videotape to provide feedback is very helpful.

**The desired outcome of Steps 5 and 6 is *initial competence*
in using both the model and the skills that make it work.**

7. **Evaluating the Learning Experience**. From time to time stop and reflect on the training process itself. Take the opportunity to express how you feel about the learning program and how you feel about your own progress. While Steps 1 through 6 deal with the task of learning the helping model and the methods and skills that make it work, Step 7 deals with developing and maintaining a learning community.

8. **Supervised Practice with Actual Clients**. Finally, when you are ready, apply what you have learned (under supervision from a trainer) to actual clients. Supervision is an extremely important part of the learning process.

The desired outcome of Steps 7 and 8 is *initial mastery*.

The program in which you are enrolled may cover only a few of these steps. Comprehensive programs for training professional helpers, however, should eventually include all steps, although they may not follow the order presented here. Of course, the best helpers continue to learn throughout their careers—from their own experiences, the experiences of their colleagues, reading, seminars, formal and informal research-- and, perhaps most importantly, from their interactions with their clients. Full mastery, therefore, is an ongoing journey, rather than a destination.

Extensive self-exploration exercises are included in this manual to help you take a good look at your own strengths and weaknesses as a helper. They will provide you with the opportunity to apply the helping model to yourself first, before trying it out on others. You can use these exercises to confirm strengths that make you effective with clients, while managing weaknesses that might stand in the way of your helping clients manage problem situations.

SHARING YOUR OWN EXPERIENCE:
BENEFITS AND CAUTIONS

A training program is an ideal opportunity to learn how to help others, as well as improve your own life. It is an opportunity to explore and deal with your own problems and unused life-enhancing opportunities (we all have them). You can do this on your own, of course. But you can also do this when you play the role of client in the training sessions. However, because sharing your own experience in the quasi-public forum of a training group entails some risks, there are laws to protect you. For instance, no one can force you to reveal yourself in public. Instructors have the legal responsibility to see that no one is exposed to harm in the training sessions. In the following pages, statements such as "Share what you have learned about yourself with a learning partner" are suggestions— not commands.

But there is a dilemma here. As an instructor, I have always taken very seriously my moral obligation to the future clients of those in my training groups. The principles are clear. Graduating or certifying incompetent trainees is simply not right. If they don't have the skills, they should not be certified. Furthermore, graduating or certifying students who have serious unmanaged problems of their own is not right. This is tricky, especially in a highly litigious society. The instructor in your training program should discuss all of these issues with you. You should know both your rights and your moral obligations.

2

As you will see, the exercises provide many opportunities to share your own experience. You have options. You can review and deal with what you learn about yourself in private, or you can share your experiences with a few people you trust outside the formal training sessions. You can discuss problems and unused opportunities with a counselor. You can be a full member of a training group where participants agree to deal with personal issues openly because both psychological and legal safeguards are in place. Or, you can opt for some combination of these strategies.

If you are in a training group where sharing personal issues to encourage personal and professional growth is the norm, consider the following guidelines.

- **Achieve a Balance Between Comfort and Risk-taking.** While sharing information about your personal life is usually challenging and at times uncomfortable, the level of your discomfort should not interfere with your learning, or become a problem in its own right. It is very difficult to learn if you are uncomfortable or feel forced. At the same time, the quality of learning improves if each learner contributes something of themselves that is real and substantial.

- **Deal With Real Issues.** Remember that you are in charge of what you reveal about yourself. If the conditions are right, you can use the training process to evaluate problems or concerns in your own life, *especially issues that influence your effectiveness as a helper*. For instance, if you tend to be an impatient person, or one who places unreasonable demands on others, you will have to examine and change this behavior if you want to become an effective helper. Or, if you are very nonassertive, this may inhibit you from helping clients challenge themselves. Another reason for using real problems or concerns when you take the role of the client is that it gives you some experience of *being* a client. Then, when you face real clients, you can appreciate some of the misgivings they might have in talking to a relative stranger about the intimate details of their lives., I would personally prefer going to a helper who has had some kind of personal experience as a client

- **Use Role Play as Needed.** An alternative to revealing your personal experiences is the use of role play. In experiential programs, you are going to be asked to act both as helper and as client in practice sessions. In the written exercises in this manual, you are asked at one time or another to play each of these roles. Role playing is not easy. It forces you to get "inside" the clients you are playing and understand their concerns as they experience them. This can help you become more empathic. However, an over-reliance on role play is not conducive to learning. It can make skills training look more like a theater class. That said, role play can allow for skills to be practiced under a wider variety of conditions than may truly exist among the members of a single training group.

- **Observe Confidentiality.** The information that is shared in the training program is not to be discussed outside of the class. Think of this as practice for when you are a counselor and your clients trust you with the stories of their lives.

GIVING FEEDBACK TO YOURSELF AND OTHERS

If you are working with a learning partner, or if you are in an experiential training group, you will be asked to give feedback to both yourself and your co-learners on how well you are learning and using helping methods and skills. Giving feedback well is an art. Here are some guidelines to help you develop that art.

1. **Keep the goal of feedback in mind.** In giving feedback, always keep in mind the generic goal of feedback-- to help the other person (or yourself) become a more effective helper. Improved performance is the desired outcome. Feedback will help you learn every stage and task in the helping model.

2. **Give positive feedback.** Tell your co-learners what they are doing well. This reinforces useful helping behaviors. "You leaned toward him and kept good eye contact, even though he became very intense and emotional. Your behavior sent the right 'I'm-still-with-you' message."

3. **Don't avoid corrective feedback.** To learn from our mistakes, we must know what they are. Corrective feedback, given in a tactful and caring way, is a powerful tool for learning. "You seem reluctant to challenge your clients. For instance, Sam [the client] didn't fulfill his contract from the last meeting and you let it go. You fidgeted when he said he didn't get around to it."

4. **Be specific.** General statements like, "I liked your style in challenging your client," or "You could have been more understanding," are not helpful. Change them to specific descriptions such as, "Your challenge was helpful because you pointed out how self-defeating her internal conversations with herself are, and you hinted at ways she could change those conversations."

5. **Focus on behavior rather than traits.** Point out what the helper does or fails to do. Do not focus on traits or use labels such as, "You showed yourself to be a leader." or "You're still a bumbler." Avoid using negative traits such as "lazy," "a slow learner," "incompetent," "manipulative," and so forth. This is just name calling and creates a negative learning climate in the group.

6. **Indicate the impact of the behavior on the client.** Feedback should help counselors-to-be interact more productively with clients. It helps, then, to indicate the impact of the helper's behavior on the client. "You interrupted the client three times in the space of about two minutes. After the third time, she spoke less intensely and switched to safer topics. She seemed to wander around."

7. **Provide hints for better performance.** Often helpers, once they receive corrective feedback, know how to change their behavior. After all, they are learning how to help through the training program. Sometimes, however, if your fellow trainee agrees with the feedback but does not know how to change his or her behavior, suggestions or hints on how to improve performance are useful. These, too, should be specific and clear. "You are having trouble providing your clients with empathy because you let them to talk too long. When they go on and on, they make so many points that you don't know which to respond to. Try interrupting your clients gently so that you can respond to key messages as they come up."

8. **Be brief.** Feedback that is both specific and brief is most helpful. Long-winded feedback is a waste of time. A helper might need feedback on a number of points. In this case, provide feedback on one or two points. Give further feedback later.

9. **Use dialogue.** Feedback is more effective if it takes place through a dialogue between the giver and receiver—a brief dialogue, of course. This gives the receiver an opportunity to clarify what the feedback giver means and to ask for suggestions if he or she needs them. A dialogue helps the receiver "own" the feedback more fully.

TRANSFERRING LEARNING FROM THE CLASSROOM TO THE REAL WORLD

The true acquisition of the helping skills is accomplished when they are at your command in everyday living-- not just when you practice them in a training program. That is why many exercises are designed to encourage you to take the skills you learn in the classroom and apply them to your daily life. The skills you will be learning are not just the skills of helping. Effective interpersonal communication, problem management, and opportunity development are essential elements of a full life.

PART ONE

LAYING THE GROUNDWORK

The centerpiece of this book is a problem-management, opportunity-development helping framework, or model, coupled with the methods and communication skills that make it work. But before you can begin to apply this model to helper-client relationships, there is groundwork to be laid. This includes:

Chapter 1

- outlining the nature and goals of helping
- becoming familiar with an outline or model of full human functioning or maturity
- exploring the challenges the helping professions face
- determining what a full curriculum for helpers might look like

Chapter 2

- describing the helping relationship and the values that should drive it

Chapter 3

- providing an overview of the helping process itself together with a case to illustrate it

Most of the exercises in these chapters are designed to help you personalize what you are learning, and apply it to your own life.

Chapter 1

INTRODUCTION TO SKILLED HELPING

Whether you become a professional helper or not, learning the model, methods, and skills of *The Skilled Helper* can help you become more effective in your interactions with yourself and others in all the social settings of life, including family, friendships, and work settings. It is important that you understand what helping is all about in order to get the most out of the exercises in this manual.

EXERCISE 1-1: UNDERSTANDING WHAT HELPING IS ALL ABOUT

1. Read the sections in *The Skilled Helper* that define what helping is all about, the goals of helping, and helping as a learning process.
2. Imagine yourself as a helper talking to a stranger you meet on a bus, train, or plane. When the stranger discovers that you are studying to be a counselor, they ask you to explain what counseling is all about.
3. On a separate piece of paper, write what you would say to the stranger. Using your own words, include a brief description of what helping is and what its goals are.
4. Share what you have written with a learning partner. Note the similarities and the differences between the two statements. To what degree do they include the main ideas outlined in the text? In light of this dialogue, how might you modify your statement?

EXERCISE 1-2: REVIEWING YOUR OWN SOCIAL-EMOTIONAL INTELLIGENCE

Below is a social-emotional intelligence framework taken from Box 1-1 in Chapter 1 of *The Skilled Helper*. Using a scale from 1-7 (1 is low, 7 is high), rate yourself on each of the items in the framework. Then follow the instructions at the end of the framework.

Box 1-1. Maturity as the Exercise of Social-Emotional Intelligence

I. Mature people are self-managers. They know themselves, are in control of themselves, and get things done.

Self-Awareness. Self-managers know themselves without becoming preoccupied with themselves.
- They know their strengths and their limitations.
- They know how they experience emotions, how they tend to express them, and what impact this expression has on themselves and others.
- They understand and accept themselves and have a realistic sense of self-worth.

Self-Control. Positive self-control (not negative self-restriction) characterizes self-managers.
- They can be trusted because it is clear to others that they have standards of honesty, integrity, and decency.
- They keep disruptive emotions and impulses in check.
- They find constructive ways of coping with stress. They take responsibility for their actions, as well as the consequences of their actions.
- They are flexible and open to new information, ideas, and ways of doing things.

Personal Agency. Self-managers have a bias toward action. They are doers rather than bystanders.
- They have life goals and pursue them.
- They are assertive without being aggressive in expressing their ideas and needs.

6

- They are interested in excellence rather than mediocrity.
- They manage problems and identify and develop opportunities.
- They persist in the face of obstacles.

II. Mature people handle relationships well. They know how to move creatively beyond themselves.

Empathy. They are aware of other people's feelings, needs, and concerns.
- They realize that empathy is inclusive; it is a two-way street.
- They listen actively and without bias.
- They seek to understand the "private logic" of others.
- They seek both to understand and to communicate their understanding.
- They often anticipate the concerns and needs of others.

Communication. They strive to communicate well with others.
- They get their points across as clearly as possible.
- They are willing to challenge and be challenged constructively.
- They are willing to negotiate and resolve disagreements.

Interpersonal Relationships. Mature people prize solid relationships with varying types and degrees of intimacy.
- They are intimate in responsible ways.
- Though capable of independent thought and action, they prize mutuality.
- They are open to influencing and being influenced, but without exercising or submitting to coercion.

III. Mature people connect with the wider world. They work at developing a "sense of the world" in which they live.

- They are not parochial but try to place their concerns in the context of what the world is really like.
- Endowed with a sense of social responsibility, they seek to be constructive members of their social groups.

As you review this framework, where do your strengths lie? What are your highest scores?

_____

_____

_____

What are some areas that need further work?

_____

_____

_____

Discuss what you have learned about yourself with a learning partner or friend.

EXERCISE 1-3: PREPARING TO PLAY THE ROLE OF CLIENT: STRENGTHS AND SOFT SPOTS IN MY LIFE

One way of preparing yourself to talk about issues that can influence the quality of your helping is to review both strengths and soft spots in your life. All of us have strengths-- things we do well-- and soft spots, or areas in which we could use some improvement. For instance, I am a decisive person, and that is a strength for me as a helper. But, in being decisive, I might push others too hard, and that consequence of my strength is a soft spot that needs improvement. In this case, a strength pushed too hard becomes a soft spot.

Here is a list of some of the kinds of problems, issues, and concerns that trainees have dealt with during training programs. As you read them, circle the ones that pertain (even if only partially) to you. Use the list to stimulate your thinking about the kinds of thinking, acting, and emotional expression that might stand in the way of your being the best possible helper.

- I'm shy. My shyness takes the form of being afraid to meet strangers and being afraid to reveal myself to others.
- I'm a fairly compliant person. Others can push me around and get away with it.
- I get angry fairly easily and let my anger spill out on others in irresponsible ways. I think my anger is often linked to not getting my own way.
- I'm a somewhat lazy person. I find it especially difficult to expend the kind of energy necessary to listen to and get involved with others.
- I'm somewhat fearful of people of the opposite sex. This is especially true if I think they are putting some kind of demand on me for closeness. I get nervous and put them off.
- I'm a rather insensitive person, or so I have been told. I'm a kind of bull-in-the-china-shop type. Not much tact.
- I'm overly controlled. I don't let my emotions show very much. Sometimes I don't even want to know what I'm feeling myself.
- I like to control others, but I like to do so in subtle ways. I want to stay in charge of interpersonal relationships at all times.
- I have a strong need to be liked by others. I seldom do anything that might offend others or that others would not approve of. I have a very strong need to be accepted.
- I have few positive feelings about myself. I put myself down in a variety of ways. I get depressed a lot.
- I never stop to examine my values. I think I hold some conflicting values. I'm not even sure why I'm interested in becoming a helper.
- I feel almost compelled to help others. It's part of my religious background. It's as if I didn't even have a choice.
- I'm sensitive and easily hurt. I think I send out messages to others that say "be careful of me."
- I'm overly dependent on others. My self-image depends too much on what others think of me.
- A number of people see me as a "difficult" person. I'm highly individualistic. I'm ready to fight if anyone imposes on my freedom.
- I'm anxious a lot of the time. I'm not even sure why. My palms sweat a lot in interpersonal situations.
- I'm somewhat irresponsible. I take too many risks, especially risks that involve others. I'm very impulsive. That's probably a nice way of saying that I lack self-control.
- I'm very stubborn. I have fairly strong opinions. I argue a lot and try to get others to see things my way. I argue about very little things.
- I don't examine myself or my behavior very much. I'm usually content with the way things are. I don't expect too much of myself or others.
- I can be sneaky in my relationships with others. I seduce people in different ways, not necessarily in a sexual way, but also with my "charm." I get them to do what I want. I'm very good at getting my own way.

8

- I like the good life. I'm pretty materialistic and I like my own comfort. I don't often go out of my way to meet the needs of others.
- I'm somewhat lonely. I don't think others like me, if they think about me at all. I spend time feeling sorry for myself.
- I'm awkward in social situations. I don't do the right thing at the right time. I don't know what others are feeling when I'm with them and I guess I seem callous.
- Others see me as "out of it" a great deal of the time. I guess I am fairly naive. Others seem to have deeper or more interesting experiences than I do. I think I've grown up too sheltered.
- I'm stingy with both money and time. I don't want to share what I have with others. I'm pretty selfish.
- I'm somewhat of a coward. I sometimes find it hard to stand up for my convictions even when I meet light opposition. It's easy to get me to retreat from my opinions.
- I hate conflict. I'm more or less a peace-at-any-price person. I run away when things heat up.
- I don't like it when others tell me I'm doing something wrong. I usually feel attacked, and I fall silent or attack back.

Using this list to "prime the pump," jot down three issues or concerns you have. Remember, choose items that might stand in the way of being an effective helper. Try to name issues that have some substance but which you would be willing to discuss in the training group or with a chosen partner.

Issue 1.

_____

_____

_____

Issue 2.

_____

_____

_____

Issue 3.

_____

_____

_____

EXERCISE 1-4: REVIEWING MY STRENGTHS

All of us have strengths, things we do well. Positive psychology suggests that the helping professions have spent too much time on clients' problems and soft spots, and not enough time helping them cultivate their strengths. Of course, an overused strength might, in some situations (including helping), turn into a soft spot.

1. In this exercise, using a separate piece of paper, jot down attitudes, skills, behaviors, and ways of expressing emotions that help you lead a fuller life (strengths). Here is what one prospective counselor had to say:

- I have lots of friends. I like people. I like being with people.
- Others can count of me.
- I've got a good part-time job; my boss likes the quality of my work.
- I'm fairly smart, but I don't go around telling people about it.
- I am very healthy. I eat right and get the kind of exercise that keeps me healthy.
- My belief in God gives me a center.
- I'm hardworking.
- I have a mind of my own; I don't give in just to please others.

2. Review your list once you are finished. Star two or three items you think will help make you a better helper.
3. Share you list with a learning partner and discuss how the starred strengths might contribute to your being a good helper.
4. Keep this list. It will help you choose issues to discuss when you play the role of client.

EXERCISE 1-5: TURNING PROBLEMS INTO OPPORTUNITIES

While on the surface the long list of issues or concerns outlined in Exercise 1-3 looks like a list of problems, it can also be seen as a list of opportunities. Take the last item, for example:

> "I don't like it when others tell me I'm doing something wrong. I usually feel attacked and I fall silent or attack back."

Once this person recognizes that they react poorly to negative or corrective feedback, they have the opportunity to change their behavior. If they work at it, there will come a time when they are able to say:

> "When people give me negative or corrective feedback, I use it as an opportunity to learn something about myself. Even when what the other person is saying is unfair, I can listen and learn. And it's an opportunity for self-discipline. I ask questions to get at the heart of what the person is saying. It's not always easy and I don't always succeed, but I keep learning about myself."

You can do the same with the three issues you chose in Exercise 1-3. Turn each problem you face into an opportunity.

Problem Turned Opportunity 1.

_____

_____

_____

Problem Turned Opportunity 2.

_____

_____

_____

Problem Turned Opportunity 3.

_____

_____

_____

EXERCISE 1-6: USING STRENGTHS TO SPOT UNUSED OPPORTUNITIES

Unused opportunities are as important as problems. This exercise helps you take a look at some unused opportunities in your own life which, if identified and pursued, will help you become a better counselor.

1. Return to the strengths section of Exercise 1-4.
2. Read your list and identify two ways you might leverage one or any combination of your strengths into opportunities that would make you a more effective helper.
3. Share the opportunities you find and talk them through with a learning partner.

Example. Sonya is a part-time student in a counseling program. She also has a job that consumes 30 hours of her time per week. She reviews her list of strengths and says, "I'm very hard working both at school and at work. I'm also very careful in how I do my work. However, I don't put the same effort into developing my social life. I could apply these two strengths in developing a fuller social life. After all, helping is about people and I need to get better at people skills." She goes on to search for a second unused opportunity. In talking her unused opportunities through with a learning partner, she comes to realize that she can use the training group as a "lab" for developing a better social life.

EXERCISE 1-7: EXPLORING THE CHALLENGES FACING THE HELPING PROFESSIONS

Chapter 1 outlines many of the challenges facing the helping professions.

1. After reading the section on challenges, answer the following questions for yourself:

- To what extent were your surprised to discover such a range of challenges facing the helping professions?
- What is your reaction to these challenges in general?
- Which ones struck you most? Which do you see as most important?
- Which ones do you think are merely pseudo-challenges or irrelevant?
- What is the upside to these challenges?
- To what extent do they make you question your interest in becoming a helper?
- In what ways do these challenges reinforce your desire to become a helper?
- How do you react to the challenge that values are not "scientific" ?

2. Share your reactions with a learning partner.

- In what ways did you and your partner agree?
- In what ways did you and your partner disagree?
- What did you learn from your discussion?

Chapter 2

THE HELPING RELATIONSHIP:
VALUES IN ACTION

It is important for you to take initiative in determining the kinds of values you want to pervade the helping process and relationship. Too often, such values are taken for granted. The position taken in Chapter Two of *The Skilled Helper* is that values should provide guidance for everything you do in your interactions with clients.

As indicated in the text, the values that should permeate your helping relationships and practice must be owned by you. Learning about values from others is very important, but mindless adoption of the values promoted by the helping professions without reflection inhibits your ability to own and practice them. You need to wrestle with your values a bit in order to make them your own. Your values must actually make a difference in your interactions with clients. They must be values-in-use and not merely espoused values. Your clients will get a feeling for your values not from what you say, but from what you do.

If possible, do Exercises 2-1 and 2-2 BEFORE reading Chapter 2.

EXERCISE 2-1: "IF I WERE A CLIENT . . .

1. Imagine yourself as a client. Think of some of the problems you have had to grapple with or are struggling with now. Then picture yourself addressing these issues with a counselor. Finally, ask yourself the following questions. Jot down words, phrases, or simple sentences in response.

What would I want the helper to be like?

_____

_____

_____

_____

How would I want to be treated?

_____

_____

_____

_____

_____

What would my responsibilities be as a client? What should I do to make helping work?

_____

_____

_____

_____

_____

2. Share your statements with a learning partner. Discuss both the similarities and the differences between your responses.

EXERCISE 2-2: YOUR PRELIMINARY STATEMENT OF VALUES

1. On a separate piece of paper, write a preliminary STATEMENT OF VALUES based on what you learned from the previous exercise. Write it in such a way that you could give it to a client, saying, "These are the values that underlie and permeate the helping process."

2. Share your statement with a learning partner. Note both the similarities and the differences. Tell each other, in light of the other's statement of values, how you would feel, and what your hopes and fears would be were you to be your learning partner's client.

3. Finally, read Chapter Two in *The Skilled Helper*. This chapter deals with values that should permeate the helping process, and offers a preliminary consideration of the relationship between client and helper. In the light of what you learn, revise your value statement and share it with a learning partner.

EXERCISE 2-3: FOUR KEY VALUES—A SELF-ASSESSMENT

The four values highlighted in Chapter 2 are respect, empathy, client empowerment and responsibility, and a bias toward action. After reading the section in the book on each value, make a personal statement as to the place this value presently has in your life. How do you put each value into action?

Respect:

_____

_____

_____

Empathy:

_____

_____

_____

Client Empowerment and Responsibility

_____

_____

_____

A Bias Toward Action

_____

_____

_____

EXERCISE 2-4: BUILDING AND MAINTAINING RELATIONSHIPS

It goes without saying that the relationship between helper and client is important. Your job is to build the kind of relationship with each client that contributes to problem-managing and opportunity-developing outcomes. This exercise asks you to look at your strengths and weaknesses in establishing and maintaining relationships in your everyday life.

1. Briefly describe your overall effectiveness in establishing solid working relationships with other people.

_____

_____

2. What do you do best in establishing and maintaining relationships?

_____

_____

3. What mistakes do you make?

_____

_____

4. What do you do to recover from mistakes?

_____

_____

14

5. What do you need to do to become better at establishing and maintaining good working relationships with others?

_____

_____

6. Discuss your responses with a learning partner.

EXERCISE 2-5: WORKING WITH DIVERSITY—A SELF ASSESSMENT

As a helper, you will meet clients that differ from you in many ways: abilities, accent, age, attractiveness, color, developmental stage, disabilities, economic status, education, ethnicity, fitness, gender, group culture, health, national origin, occupation, personal culture, personality variables, politics, problem type, religion, sexual orientation, social status, and values—to name some of the major categories. Your ability to work with clients that are different from you is one of the major requirements for being an effective helper. On a scale from 1-7 (7= a very high score) rate yourself on the following statements.

- I enjoy meeting and interacting with people who are different from me.
- I would not like to live (don't like living) in a homogeneous culture.
- Even when my initial reactions to people are negative, I try to understand rather than judge them.
- I fully understand that people (including myself) are not perfect and I make reasonable allowances for this.
- I probably have blind spots when it comes to the way I treat people, but I'd like to know what they are so that I could deal with them.
- I see culture as a two-way street: I want to understand and respect the cultural differences of others, but I also expect them to do the same for me.
- I realize that my understanding of other people's cultures will always be partial and fallible.
- I want to understand others in the key diversity and cultural contexts of their lives.
- Because of the richness of human diversity, when I meet and deal with others I am always a learner.
- I think that it is important for me to understand whatever diversity and cultural biases (racism, sexism, and ageism, for example) I have and to deal with them.
- Although I realize that differences can be a cause of conflict, I believe that differences can also be a source of human enrichment.

Consider your high scores and your low scores. Share key insights about yourself with a friend or learning partner.

Chapter 3

AN OVERVIEW OF THE HELPING PROCESS

In Chapter 1, you were asked to explain the overall goals of counseling to another person. Chapter 3 goes into more detail about the helping process itself. Do this exercise before reading Chapter 3 or the outline of the helping process below.

EXERCISE 3-1: WHAT DO PEOPLE THINK COUNSELING INVOLVES?

Follow the process outlined here before reading Chapter 3 of the book.

1. Interview three or four friends, relatives, or colleagues. Ask them what they think counseling is.
2. Get more than just a definition. Help them expand on whatever definition or description they provide.
3. Write a summary of the main facets in their definitions and descriptions.
4. Meet with two or three of your fellow learners and share what you've discovered.
5. What definitions/descriptions of counseling emerge from your discussion? What are the themes? What are the variations?
6. After studying Chapter 3, discuss with the same group how the "public" sees counseling,
7. Finally, discuss the importance of sharing with clients what counseling is, what the helper's role and responsibilities are, and what the client's role and responsibilities are.

EXERCISE 3-2: WHAT KIND OF PROBLEM-MANAGER AND OPPORTUNITY-DEVELOPER AM I?—A SELF-ASSESSMENT

As a counselor, you are going to be helping clients manage problem situations and develop unused or underused opportunities. It would be odd if you were not utilizing these same skills in managing your own life. Here is an opportunity to use the helping model, or framework, as a self-assessment tool. Helpers can and should become better problem-managers throughout their lives. Rate yourself on each of following problem-managing qualities on a scale from 1-7 (with 7 as a high score).

Stage I-Task 1 Behavior
- I spot problem situations as they begin to develop.
- I have a "nip them in the bud" mentality.
- I act on the "prevention is better than cure" principle.
- I'm always on the watch for life-enhancing opportunities.
- I analyze both problem situations clearly; I quickly grasp what's going on.
- I quickly analyze the potential of any given opportunity.
- I apply these behaviors to the rest of problem management—goal setting, strategy development, and implementation.

Stage I-Task 2 Behavior
- I am open to getting feedback from others.
- I find ways of getting in touch with my blind spots.
- I am open to exploring my blind spots with others.
- I translate blind spots into useful, life-enhancing new perspectives.
- I understand that I can have blind spots during any stage or task of the helping process.
- I am a vigilant person.

16

Stage I-Task 3 Behavior

- I know how to separate the important issues of life from the trivial.
- I know the difference between what is critical and what is merely useful.
- In both problem management and opportunity development I focus on the bigger issues of life.
- I do not let myself become a victim of the trivial things of life.

Stage II-Task 1 Behavior

- I'm open to problem-managing and opportunity-developing possibilities.
- The psychology of hope is important for me.
- I brainstorm about the future.
- The concept of "possible selves" is important for me.
- I look for ways to develop my creativity.

Stage II-Task 2 Behavior

- Goals play an important role in my life.
- Even though I am goal-driven, I am still flexible.
- My goals have the kind of focus and clarity needed to drive action.
- I appreciate all kinds of goals—emerging, adaptive, and coping.
- I tailor goals to the kind of problem situation or opportunity at hand.

Stage II-Task 3 Behavior

- I have a practical understanding of the dynamics of incentives, commitment, and motivation.
- The goals I set are both useful and appealing.
- I take my commitments—both to myself and to others—seriously.

Stage III Behavior

- I know that there are usually a number of paths to a goal.
- I explore these paths diligently.
- I choose strategies that reflect my goal-meeting style and are appropriate to the circumstances.
- Whenever required, I turn the strategies I choose into a step-by-step plan.
- My plans are living realities, not just words on a page.

Action Arrow Behavior

- I am a doer, not just a dreamer.
- As soon as I start to think about a problem situation or an opportunity, I begin engaging in actions that move me toward substantial outcomes.
- The stages and tasks outlined above make me think of things I can do.
- I am open to changing my cognitive behavior, my external behavior, and/or the ways I express my emotions.
- My actions are usually focused, not just actions for the sake of action.

General Problem Management Behavior

- The behaviors outlined above constitute a mentality and a way of approaching life.
- The behaviors outlined are often implicit rather than explicit; I am not a goal junkie.
- The framework is in my bones; it is part of my lifestyle.

EXERCISE 3-4: PREPARING TO TELL YOUR STORY—PROBLEM SITUATIONS

All helpers have problems or issues that can diminish their effectiveness. Knowing what yours are is an important part of training. Acknowledging this, as well as doing the following sentence-completion exercise, will help you to choose issues that you would like to and are willing to discuss. Do these sentence-completion exercises quickly.

1. One of my biggest problems is _____.

2. I'm quite concerned about _____.

3. Something I fail to do that gets me into trouble is _____.

4. The social setting of life I find most troublesome is _____.

5. The most frequent negative feelings in my life are _____.

6. These negative feelings take place when _____.

7. The person I have most trouble with is _____.

8. What I find most troublesome in this relationship is _____.

9. Life would be better if _____.

10. I tend to shortchange myself when I _____.

11. I don't cope very well with _____.

12. I get anxious when _____.

13. A value I fail to put into practice is _____.

14. I'm afraid to _____.

15. I wish I _____.

16. I wish I didn't _____.

17. What others dislike most about me is _____.

18. I don't seem to have the skills I need in order to _____.

19. A problem that keeps coming back is _____.

20. If I could change just one thing in myself it would be _____.

Once you have finished, place a check-mark next to the issues that you might be willing to discuss with a friend, partner, or in the training group. Then circle the checked issues you believe are most closely related to becoming an effective helper. Finally, with a learning partner, discuss the items you have both checked and circled.

All in all, what kind of problem manager are you? Where do your strengths lie? Your weaknesses?

_____

_____

_____

EXERCISE 3-5: PREPARING TO TELL YOUR STORY—UNUSED OPPORTUNITIES

This sentence-completion exercise mirrors the preceding one, but deals with unused or underused strengths and opportunities rather than problems. Undeveloped strengths and opportunities can affect the quality of your helping.

1. I could be better at _____.

2. Those who know me know that, when I want to, I can _____.

3. I've always wanted to _____.

4. I regret that I haven't _____.

5. When I'm at my best I _____.

6. I'm glad when I _____.

7. Those who know me know that I can _____.

8. A value that I try hard to practice is _____.

9. When I'm at my best with people I _____.

10. I can't always count on myself to _____.

11. Something I'm handling better this year than last is _____.

12. If I were managing my life better I would _____.

13. A recent problem I haven't handled as well as I might is _____.

14. One goal I'm presently working toward is _____.

15. I aspire to _____.

16. I think that I have the guts to _____.

17. An example of some unfinished business I have is _____.

18. When I think of my maturity I _____.

19. One way in which I could be more dependable is _____.

20. I communicate most effectively with others when _____.

Once you have finished, place a check-mark next to the unused opportunities that you might be willing to discuss with a training group. Then circle the checked issues you believe are most closely related to becoming an effective helper. Finally, with a learning partner, discuss the items that you have both checked and circled.

All in all what kind of opportunity-identifier and developer are you? Where do your strengths lie? Your weaknesses?

_____

_____

_____

PART TWO

STAGE I OF THE HELPING PROCESS & THE SKILLS OF THERAPEUTIC DIALOGUE

Part Two, which includes Chapters 4-9, focuses on the three tasks of Stage I : Helping clients tell their stories. It will also address the communication skills you will need throughout the entire helping process. These communication skills are the essential components of the *therapeutic dialogue* between helper and client. As we shall see, dialogue is at the heart of the helping process, enabling clients and helpers to engage collaboratively and productively. Chapter 4 includes an overview of the first task of Stage I. There is also a section on how to address reluctant and resistant clients.

In a nutshell, Part Two will help you to engage in the three interrelated tasks of Stage I; to help clients tell their *stories* (Task 1), to help clients tell their *real* stories without fudging (Task 2), and to help clients tell the *right* stories and discuss issues that will make a difference in their lives (Task 3). Part Two will also provide you with the communication skills needed to successfully accomplish tasks in each of the three Stages.

Chapter 4

STAGE I: THE CURRENT PICTURE

Clients come to helpers because they need help managing their lives more effectively. Stage I illustrates three ways in which counselors can help clients understand themselves, their problem situations, and their unused opportunities. There are three interrelated tasks in Stage I:

Task 1: Stories. Help clients tell their stories, share their points of view, explore their decisions, and clarify their intentions and emotions.

Task 2: New Perspectives. Help clients identify and move beyond blind spots to new perspectives that will help them manage their problem situations and opportunities more effectively.

Task 3: Value. Help clients work on issues that will most likely make a difference in their lives.

Chapter 4 focuses on Task 1: helping clients tell their stories as clearly as possible. Since these three tasks are interrelated and apply to all the stages and tasks of the helping process, clarity is always necessary. Remember that the principles embedded in these three tasks are not restricted to Stage I-- first of all, clients rarely tell their full stories at the beginning of the helping process. Often the story "leaks out" over time. Secondly, new perspectives can be helpful at any stage of the helping process. New perspectives are useful in choosing goals, developing strategies for implementing goals, and in the implementation process itself. Lastly, value in the broadest sense deals with the "economics" of helping. Choosing the right problem or opportunity to work on is just the beginning. Choosing the right goals, strategies, and resources for implementation are also part of the picture. Chapter 4 has two sections. Section I deals with the first task of Stage I, and. Section II addresses client reluctance, resistance, and resilience.

SECTION I.
STAGE I: TASK 1:
PARTNER WITH CLIENTS IN HELPING THEM TELL THEIR STORIES

Task 1 deals with helping clients tell their stories-- discussing problem situations, anxieties, and concerns that bring them (or get them sent) to the helper in the first place, as well as identifying and discussing unused opportunities. Read Section I of Chapter 4 in the text before doing the exercises in this section.

We begin with an exercise in self-exploration, focused on some normative developmental tasks. Many of the clients you see will be struggling with developmental concerns like those you have struggled with or are currently working on. Like you, they will have both strengths and "soft spots" in coping with the developmental tasks of life.

EXERCISE 4-1: REVIEWING SOME BASIC DEVELOPMENTAL TASKS

In this exercise you are asked to consider your experience with ten major developmental tasks of life.

1. First, reflect on your own experience in these developmental areas. What you discover will help you play the role of the client more effectively in later exercises.

1. **COMPETENCE. What do I do well?** To what extent do I see myself as a person who is capable of getting things done? How effective am I in gathering the resources needed to accomplish the goals I set for myself? In what areas of life do I excel? In what areas of life would I want to be more competent than I am?

Strengths	**Soft Spots**
_____	_____
_____	_____
_____	_____
_____	_____
_____	_____

2. **AUTONOMY: How well do I make it on my own**? How effectively do I avoid being either overly dependent or too independent? To what degree am I reasonably interdependent in my work and social life? When I need help, how easy do I find it to ask for it? In what social settings do I find myself most dependent? Counterdependent? Independent? Interdependent?

Strengths	**Soft Spots**
_____	_____
_____	_____
_____	_____

3. **VALUES: What do I believe in and prize?** What are my principal values? To what degree do I pursue reasonable development in my value system? How well do I put my values into practice? What value conflicts do I face? In what social settings do I pursue the values that are most important to me?

Strengths	**Soft Spots**
_____	_____
_____	_____
_____	_____
_____	_____

4. **IDENTITY: Who am I in this world?** To what extent do I have a sense of who I am and the direction I'm going in life? What kind of congruence exists between the way I see myself and the way others see me? What kind of "center" gives meaning to my life? In what social settings do I have my best feelings for who I am? In what social settings do I tend to lose a sense of who I am? In what ways am I confused or dissatisfied with who I am?

Strengths   **Soft Spots**

_____   _____

_____   _____

_____   _____

_____   _____

_____   _____

5. INTIMACY. What are my closer relationships like? What kinds of closeness do I have with others? To what extent are there degrees of closeness in my life — from acquaintances to friends to intimates? What is my life in my peer group like? How well do I get along with others? What concerns do I have about my interpersonal life?

Strengths   **Soft Spots**

_____   _____

_____   _____

_____   _____

_____   _____

6. SEXUALITY. Who am I as a sexual person? To what degree am I satisfied with my sexual identity, my sexual preferences, and my sexual behavior? How do I handle my sexual needs and wants? What social settings influence the ways I act sexually?

Strengths   **Soft Spots**

_____   _____

_____   _____

_____   _____

_____   _____

7. LOVE, MARRIAGE, FAMILY. What are my deeper commitments like? What is my marriage or intimate relationship like? How do I relate to family and relatives, however defined? How do I feel about the quality of my family or friendship-circle life? If not married, in what ways do I look forward to marriage or to some deeper stable relationship? What misgivings do I have about such a relationship?

25

Strengths	Soft Spots
_____	_____
_____	_____
_____	_____
_____	_____
_____	_____

8. **CAREER. What is the place of work in my life?** How do I feel about the way I am preparing myself for a career or the career I am currently pursuing? What do I get out of work? What am I like in the workplace? How does the workplace affect me? What impact do I have there?

Strengths	Soft Spots
_____	_____
_____	_____
_____	_____
_____	_____
_____	_____

9. **INVESTMENT IN THE WIDER COMMUNITY**. How big is my world? How do I invest myself in the world outside of friends, work, and the family? What is my neighborhood or community like? What kind of concerns do I have about community, civic, political, social issues? In what ways am I optimistic about the world? In what ways am I cynical?

Strengths	Soft Spots
_____	_____
_____	_____
_____	_____
_____	_____
_____	_____

26

10. **LEISURE. What do I do with my free time?** Do I feel that I have sufficient free time? How do I use my leisure? What do I get out of it? In what social settings do I spend my free time?

Strengths   Soft Spots

_____   _____

_____   _____

_____   _____

_____   _____

_____   _____

2. Choose two or three areas you think need some attention. What are your choices? What are the issues? Why are they important?

EXERCISE 4-2: HELPING ONE ANOTHER TELL YOUR STORIES

In this exercise you have the opportunity of playing three successive roles: the client, the helper, and the observer.

1. Form groups of three.
2. In each group there are three roles: client, helper, and observer. You will each play all three roles once. Quickly decide who will play what role to begin.
3. Before the dialogue between client and helper begins, the trainee in the client role gives a brief summary of the problem situation or unused opportunity he or she is going to explore. Choose a developmental issue or any other issue you are willing to explore. Remember the rules mentioned in the Introduction. No one should be forced to talk about themselves unless they do so freely.

Example: Enrico, the trainee playing the role of client, gives the following summary: "I'm going to discuss a problem I'm having with my parents. I'm an only child and they simply won't let go. Don't get me wrong. They're very nice about it. But I'm 21. We're not a wealthy family. So, for the time being, I have to live at home. Also I believe that becoming my own person is directly related to becoming an effective helper."

1. The helper begins by saying: "Against that background, Enrico, what would you like to focus on today?"
2. Then client and helper engage in a helping dialogue. The helper helps the client tell his or her story clearly and in useful detail.
3. After about six minutes the observer, in timekeeper mode, stops the dialogue.
4. The client gives the helper feedback on how effectively he or she has aided the storytelling process.
5. Because giving feedback to your fellow learners is an important part of the training program, review the section in the Introduction on guidelines for giving effective feedback.
6. Then the observer shares his or her observations.
7. The process continues until each member has had the opportunity to play each role.
8. After each group member has played each role, discuss what you have learned. What did helpers do that enabled clients to speak more easily and clearly?

EXERCISE 4-3: TALKING PRODUCTIVELY ABOUT THE PAST

Clients will-- out of necessity-- talk about what has happened to them in the past. In some cases the client uses the past to add depth to their story or to supply relative context. However, at other times, talking about the past without an effort to make a link to what is presently happening in the client's life can inhibit the helping process. Read the section in *The Skilled Helper* on how to make talking about the past productive.

1. Pick a partner to do this exercise either inside or outside the classroom.
2. In this exercise listen to your partner discuss a past event in his or her life.
3. Next, invite your partner to draw from the past event some implications for the present. Base your invitation on the principles outlined in Chapter 4.
4. When you have finished discuss the dialogue with your partner. How did you help your partner connect the past to the present? What else could you have done to help your partner see the implications of the past for the present?

Example: Imagine yourself talking with a 33-year-old man who has just started dating someone seriously. He is reflecting on a past relationship that didn't work out. In your dialogue with him, you learn the following: "I guess I didn't realize it then, but she was a really great woman, one of the best I will ever know. We fell for each other and everything was great. Then we had this fight. I took a very hard stand and made it impossible for her. I was really stubborn. Angry, too. Then I began seeing other women. Of course, she found out. I might have salvaged it even then if I had gone to her and said what I really felt. But, I just did what I always do — called up some buddies and went drinking."

Here are some possible responses that might help the client to constructively connect the past with the present..

Response A: "I wonder how you might use what you've learned from this past experience to prevent something similar happening in your new relationship."
Response B: "It sounds like you learned — somewhat painfully — something very useful from that relationship. What did you learn?"
Response C: "If you had that relationship to do over again, what would you do differently?"

EXERCISE 4-4: LINKING STORYTELLING TO ACTION

As noted in the text, clients need to engage in "little" actions throughout the helping process that get them moving in the right direction even before formal goals and action strategies are established. These little actions are signs of clients' commitment to the process of constructive change. In this exercise you are asked to put yourself in the place of the client and come up with some possible actions that the client might take to move forward in some way. It is not that you are going to tell the client what to do, but helping clients adopt an action-orientation to their problem situations is central to helping.

1. Read each of the stories outlined below.
2. If you were the client, what are two common-sense actions you might consider taking at this stage in order to move forward? Even if "big" actions seem premature, what "little" actions might you take? Indicate the reason for each possible action.
3. Share your findings with a learning partner and get feedback on the actions you propose and the reasoning behind them.

Example. A seventh-grade boy talking to a teacher he trusts (all this is said in a halting voice and he does not look at the teacher): "Something happened yesterday that's bothering me a lot. I was looking out the window after school. It was late. I saw two of the guys, the bullies, beating up on one of my best friends. I was afraid to go down there. . . . A coward. . . . I didn't tell anyone, I didn't do anything."

> **One possible action**: First, the boy might talk the incident through with his friend and perhaps apologize for letting him down
> **Reason**: to reestablish relationship with his friend and to show basic decency.
> **Another possible action**: Second, the boy might figure out how he wants to handle similar situations in the future.
> **Reason**: to learn something useful from this painful experience.

1. A young woman talking to a counselor in a center for battered women. She has been seen by other counselors on two different occasions: "This is the third time he's beaten me up. I didn't come before because I still can't believe it! We're married only a year. After we got married, he began ordering me around in ways he never did before we got married. He'd get furious if I questioned him. Then he began shoving me if I didn't do what he wanted fast or right. And I just let him do it! I just let him do it! (She breaks down and sobs.) And now three beatings in less than two months. Oh God, what's happened?"

Two possible client actions and a reason for each.

_____

_____

_____

_____

_____

_____

2. A 70-year-old man, arrested for stealing funds from the company where he has worked for 25 years, talking to an assigned social worker. This is their second session: "To tell you the truth, it's probably a good thing I've been caught. I've been stealing on and off for the last five or six years. It's been a game. It soaked up my energies, my attention, distracted me from thinking about getting old. Now I'm saying to myself: 'You old fool, what you are running from?' You're probably thinking: 'It's about time, old guy.'"

Two possible client actions and a reason for each.

_____

_____

_____

_____

_____

_____

3. A 15-year-old girl, talking to a psychologist at a time when her parents are involved in a divorce case: "I still want to do something to help, but I can't. I just can't! They won't let me. When they would fight and get real mean and were screaming at each other, I'd run and try to get in between them. One or the other would push me away. They wouldn't pay any attention to me at all. They're still pushing me away. They don't care how I think or feel or what happens to me! My mother tells me that kids should stay out of things like this."

Two possible client actions and a reason for each.

_____

_____

_____

_____

_____

_____

4. A 24-year-old man is talking to his mentor, a senior partner in the company. A department head has offered him a managerial job in her area. In this job he will be leading an important project. He says, "I didn't think this kind of opportunity would come so soon. But I've decided to say yes. I'm a bit worried, sure, but I figure I was asked because they think I can do the job. You know, just three years ago I was studying this stuff in school and now I'm going to have a budget, people reporting to me, and a chance to make a mark. I have a pretty clear idea about what I'm going to do when I get started next week. Most of all I want to get off on the right foot with the people."

Two possible client actions and a reason for each.

_____

_____

_____

_____

_____

_____

SECTION II.
HELP CLIENTS MANAGE RELUCTANCE AND RESISTANCE

In order to recognize reluctance and resistance in clients, it helps to first be able to recognize the same tendencies within yourself and learn how to manage them in your own life. So some of these exercises will help you look at your own reluctance and resistance. Furthermore, most clients are more resilient than they first seem to be. Helping them tap into their hidden or unused resources will help them move forward. Read Section II in Chapter 4 in *The Skilled Helper* before doing the following exercises.

EXERCISE 4-5: MY OWN RELUCTANCE

Since living more effectively requires hard work, is sometimes painful, and entails risk, all of us express reluctance from time to time. This exercise targets your own reluctance to become more than you are.

1. Review the developmental challenges in your life that you identified in EXERCISE 4-1.
2. Identify three areas or ways in which you have been reluctant to grow or change. Spell them out in some detail. Note the difference between reluctance and resistance.

Example. Michaela, a 19-year-old first-year student in clinical psychology, has this to say about developing a sense of responsibility: "The more I learn about the world, the more I see how comparatively easy life has been for me. Everything has been given to me. I'm a product of the American dream. I had an easy time in high school, but the first months of graduate school have been jarring. I'm coming to realize that no one is going to give me anything. As I look down the road I see that grad school more accurately reflects the real world. Without getting into useless guilt trips, I realize I'm spoiled. I also realize that old habits do not die easily. I still expect things to be handed to me on a platter. I'm reluctant to face up to the realities of imposed schedules, competition, demanding work, and budgets. It is very difficult for me to accept that some of my learning is gong to come from making mistakes, being challenged, and having to do things over." Now identify three areas of your own reluctance to grow.

First area of reluctance.

_____

_____

_____

Second area of reluctance.

_____

_____

_____

Third area of reluctance.

_____

_____

_____

3. As you share these instances of personal reluctance with a learning partner, see if you can identify any underlying themes that run through all three areas.

_____

_____

_____

4. Choose one of the areas or the theme that runs through all three and indicate what you might do in order to overcome your reluctance.

_____

_____

_____

_____

_____

EXERCISE 4-6: MY OWN EXPERIENCE OF RESISTANCE

Recall that resistance is a reaction to someone's trying to force you to change when you don't want to. The other person may or may not be pressuring you, but you *feel* pressured. You may even think that the other person has a point, but the fact that he or she is *telling* you what to do causes you to react negatively rather than respond constructively.

1. Recall two instances when you resisted growth because you felt you were being forced into it.
2. Indicate how you could have transformed that negative experience into a positive one for yourself.

Example. Manfred, a 24-year-old graduate student in social work, recalls being badgered by a former girlfriend: "Elise was out to reform me. It's not that I was not in need of reform during my college years. But the way she went about it turned me off. For instance, I drank too much. But she lectured me in private. Then she embarrassed me in front of my friends—her way of getting my attention. Before going out to major events like the homecoming dance, she laid down rules and made me promise to keep them. Otherwise she wouldn't go out with me. Things like that."

When asked how that negative experience could have been turned into a positive one, he had this to say: "The substance of what she was saying actually made sense. But I would have done better if she had backed off a bit, quite a bit at times. For instance, instead of laying down rules to prevent disasters, she could have quietly left the scene if they did occur. If she had said that she was no longer having fun and was going home and then just did it a couple of times, I think that would have been a real wake-up call for me. I could have said all of this to her, but I never did. I just reacted."

One instance of resistance in your own life.

_____

_____

_____

Indicate how that negative experience could have become a positive one for you.

_____

_____

_____

_____

A second instance of resistance in your on life.

_____

_____

_____

Indicate how that negative experience could have become a positive one for you.

_____

_____

_____

_____

EXERCISE 4-7: YOUR OWN WAYS OF CAUSING RESISTANCE

Resistance in clients is often prompted by real or perceived coercion on the part of the helper. As a helper, you may not think that you are putting any undue pressure on a client, but he or she might feel that you are. That is, you don't intend to be coercive but nevertheless you come across that way. For example, let's assume that you are very open about yourself. That's the way you grew up. However, when you place a demand on others for higher levels of self-disclosure than they are ready for, they may clam up.

Therefore, there is much to be gained from knowing what aspects of your own style might come across to others as subtly, or not so subtly, coercive.

Example: Marisol is a 42-year-old experienced counselor working in a health-care facility. Recently, she transferred within the hospital from the emergency room, where she handled individuals and families in crisis, to the slower-paced oncology unit.. She talks about how her emergency room style has caused problems in the oncology unit: "Down in the ER we had to work fast and I got good at making things happen. Now I'm in oncology and I'm getting on people's nerves. I feel like I'm out of synch. I want to help both staff and patients make decisions but our timetables are all different. I'm moving fast, you know, the way I did in the emergency room. And I want people to move at my pace. I'm scaring some people. Others are grateful for my assertiveness and candor. I need to tailor my style to the unit, to the different situations, and to different people. I have to pay close attention to what people need and then pace myself."

1. Jot down some aspects of your interpersonal style that might be seen as coercive by others. If you see none, don't invent any. Just say so.
2. Discuss with your partner what aspects of your helping style could be a problem for certain clients.
3. Finally, talk about ways of changing or softening your style.

Coercive aspects of your style.

_____

_____

_____

_____

Changes that would help.

_____

_____

_____

_____

EXERCISE 4-8: REVIEWING YOUR OWN RESILIENCE

Chapter 4 ends with a section on the importance of resilience. Getting in touch with your own resilience will help you spot resilience resources in your clients. Resilience means getting back on your feet after some kind of crisis or failure. It also means moving ahead in spite of continuing obstacles or opposition.

1. Think of a couple of instances of crisis or failure in your life that you handled well.
2. Describe what you did to get back on your feet.

Example. Nathan came back from a business trip to Southeast Asia extremely tired, and with severe pains in his gut and extreme diarrhea. Doctors first diagnosed diverticulitis and were considering surgery. Two days later they found that he had giardiasis. Bugs picked up during his travels were causing havoc in his gut. To make things worse, the doctors were reluctant to give up the diverticulitis diagnosis and the surgical team kept circling him. Nathan was in the hospital for ten days, an eternity in these days of managed care. The third day of his stay, he decided that he had to keep going. He put himself on a schedule. He got up at his usual time. He shaved. He put himself on a reading schedule. When he was too miserable to read, he walked the corridors, endlessly dragging an intravenous feeding device along with him. He engaged the head of the surgical team in a discussion at the heart of which was this question: "If we know for sure that I have a bad case of giardiasis and if the symptoms of that disease are pretty much the same as a severe bout of *suspected* diverticulitis, why am I getting so much attention from the surgical team?" Nathan's motto was: "I'm not going to let this do me in physically and especially psychologically." He used the resources he used in his business life—a fighting spirit, lack of self-pity, self-discipline, logic, and planning—to remain resilient in the face of the disease and medical fumbling.

Describe one crisis or failure that you feel you handled well.

_____

_____

Describe in some detail what you did to get back on your feet.

_____

_____

_____

_____

Describe a second crisis or failure that you feel you handled well.

_____

_____

Describe in some detail what you did to get back on your feet.

_____

_____

_____

_____

3. With a learning partner share answers to the question, "How resilient are you?"

EXERCISE 4-9: REVIEWING ANOTHER PERSON'S RESILIENCE

You have probably witnessed any number of examples of resilience among family, friends, acquaintances, work colleagues, fellow church members, and so forth. Not necessarily heroic resilience, but the kind of resilience many people are capable of.

1. Think of a couple of instances of substantial resilience you have witnessed.
2. Describe what these people did to get back on their feet.

Case #1.

_____

_____

In what ways did this person manifest his or her resilience?

_____

_____

_____

_____

Case #2.

_____

_____

In what ways did this person manifest his or her resilience?

_____

_____

_____

3. What does this say about the possible resilience of clients you will be seeing? Share your thoughts with a learning partner.

Chapter 5

THE COMMUNICATION SKILLS OF THERAPEUTIC DIALOGUE: THE SKILLS OF TUNING IN AND ACTIVELY LISTENING

Much of helping takes place through a dialogue between client and helper. The purpose of this dialogue is to serve the overall problem-management and opportunity-development goals of the helping process, and thus, the quality of the dialogue is critical.

THE IMPORTANCE OF DIALOGUE

As useful as dialogue might be in human communication, it is not that common—at least in its fullest form. Helpers need to become skilled in dialogue, as well as in helping clients engage in dialogue.

EXERCISE 5-1: UNDERSTANDING AND PERSONALIZING THE FOUR REQUIREMENTS OF TRUE DIALOGUE

1. Read the section in Chapter 5 that describes the requirements of what a true dialogue should look like.
2. On a scale from 1-7 (7 is a high score), rate yourself on each requirement.
3. Describe what happens in a conversation when each of these requirements is missing.
4. Compare and discuss your ratings and responses with those of a learning partner.

Turn-Taking

Rating ( )_____

_____

Connecting

Rating ( )_____

_____

_____

Mutual Influence

Rating ( )_____

_____

_____

Co-Creating Outcomes

Rating ( )_____

_____

_____

EXERCISE 5-2: DIALOGUE IN EVERYDAY LIFE

Not all conversations need to be true dialogues. However, most conversations would probably be more productive if they were conducted in the *spirit* of true dialogue. Often enough, there are times when dialogue in its fullest sense would make a big difference. In this exercise you are asked to spend the next week or two observing yourself and others during conversations. See how much time you and your conversational partners spend in dialogue, especially when important topics are being discussed.

1. On a scale from 1-7, rate yourself on how good you think you are at dialogue in everyday life. _____
2. Over the next week or two, observe yourself and your conversational partners during conversations.
3. How well did you do in dialogue, especially when true dialogue would have added value to the conversation?
4. At the end of this experiment, re-rate yourself on dialogue. What's your new score? _____
5. What could you do to improve your ability to engage in dialogue?
6. Discuss what you have learned with a partner.

ATTENDING: VISIBLY TUNING IN

Your posture, gestures, facial expressions, and voice all send nonverbal messages to your clients. The purpose of the exercises in this section is to make you aware of these varying kinds of nonverbal messages, and how to use nonverbal behavior to make contact and communicate with them. It is important that what you say verbally is reinforced, rather than muddled or contradicted, by your nonverbal messages. Before doing these exercises, read the rest of Chapter 5 in *The Skilled Helper*. There are two main points that you should take away from this chapter. First, learn to use your posture, gestures, facial expressions, and voice to *send messages* you want to clients to hear, such as, "I'm listening to you very carefully" or "I know what you're saying is difficult for you, but I'm with you." Second, become aware of the messages your clients are sending to you through *their* nonverbal behaviors.

EXERCISE 5-3: VISIBLY TUNING IN TO OTHERS IN EVERYDAY CONVERSATIONS

This is another exercise you will do outside of the training group, in your everyday life. Observe the way you attend to others for a week or two — at home, with friends, at school, and at work. Observe the quality of your presence to others when you engage in conversations with them. Of course, even being asked to "watch yourself" will induce changes in your behavior; you will probably tune in more effectively than you ordinarily do. The purpose of this exercise is to sensitize you to attending to your own behaviors in general, and to get some idea of what your day-to-day attending style looks like. First, read about the skills of tuning in, or attending, in *The Skilled Helper*.

1. On a scale from 1-7, rate yourself on how effectively you think you pay attention or tune in to others in your everyday conversations. _____

2. Without becoming preoccupied with every little behavior, watch yourself for a week as you tune in (or fail to tune in) to others in your everyday conversations.

3. What are you like when you are with others, especially in serious situations? What do you do well?

4. Re-rate yourself _____

5. Read the following example, and then write your own summary of what you have learned about yourself.

Example. Here is what one trainee wrote: "I found myself tuning in better to people I like. When I was with someone neutral, I found that my eyes and my mind would wander. I also found that it's easier for me to tune in to others when I'm rested and alert. When I'm physically uncomfortable or tired, I don't put in much effort to tune in. I was unpleasantly surprised to find out how easily distracted I am. Often I was there, but I really wasn't there. However, simply by paying attention to tuning in skills, I was with others more fully, even with neutral people."

_____

_____

_____

_____

_____

_____

6. What are some things you can you do to tune in more effectively to others?

_____

_____

_____

_____

EXERCISE 5-4: OBSERVING AND GIVING FEEDBACK ON QUALITY OF PRESENCE

In the training sessions, make sure that your nonverbal behavior is helping you work effectively with others, as well as sending the messages you want to send. This exercise, then, pertains to the entire duration of the training program. You are asked to give ongoing feedback-- both to yourself and to other members of the training group-- on the quality of your presence as you interact, learn, and practice helping skills. Recall especially the basic elements of visibly tuning in, summarized by the acronym *SOLER*. Here is a checklist to help you provide that feedback to yourself and your fellow helpers. Review criteria for giving useful feedback (this can be found in the Introduction to the Exercises).

- How effectively are you using postural cues to indicate your willingness to work with the client?

39

- In what ways do you distract clients and observers from the task at hand (for example, by fidgeting)?
- How flexible are you when engaging in *SOLER* behaviors? To what degree do these behaviors help you be with the client?
- How natural is it for you to tune in to the client? What indications are there that you are not being yourself?
- To what degree is your psychological presence reflected in your physical presence?
- What are you like when you miss the mark?
- What are you like when you are at your best?

What do you have to do to become more effectively present for your clients, both physically and psychologically?

_____

_____

_____

_____

Of course, more important than nonverbal behavior is the <u>total quality</u> of your being with and working with your clients. Your posture and nonverbal behavior are a part of your presence, but there is more to presence than *SOLER* activities. There are also the values and spirit you bring to your encounters with your clients.

ACTIVE LISTENING

Read the sections on active listening in Chapter 5 of *The Skilled Helper*. Effective helpers are active listeners. When you listen to clients, you listen to their stories. Some of the elements of these stories are:

- their **experiences**: what they see as happening *to* them;
- their **behaviors**: what they do or fail to do;
- their **affect**: the feelings and emotions that arise from their experiences and behaviors;
- the **core messages** in their stories;
- their **points of view** about key topics, including the reasons for their points of view and the implications for holding any given point of view;
- the **decisions** they are making, together with the reasons for making them, as well as the implications or possible consequences of their decisions;
- their **intentions**: the goals they are pursuing and the actions they intend to engage in;
- the wider **context** of their stories, points of view, decisions, and intentions;
- the **slant** they might give to the aforementioned elements..

Helpers need to listen carefully in order to respond with both understanding and, as we shall see later, some sort of challenge. Let's start by having you listen to yourself.

EXERCISE 5-5: LISTENING TO YOURSELF AS A PROBLEM SOLVER

In this exercise you are asked to "listen to" a problem situation or unused opportunity you have dealt with successfully. Retell the story to yourself in summary form.

What was the issue?

_____

_____

_____

What were your key experiences? What happened to you?

_____

_____

_____

What actions on your part contributed to the problem situation?

_____

_____

_____

Which of your points of view were involved?

_____

_____

_____

What decisions were involved?

_____

_____

_____

What emotions did you experience and express?

_____

_____

_____

What did you do to cope successfully with the problem or develop the opportunity?

_____

_____

_____

Share what you have written with a learning partner. Discuss how well both of you have listened to yourselves in outlining the core messages and issues of the problem or opportunity.

EXERCISE 5-6: LISTENING TO YOUR OWN FEELINGS AND EMOTIONS

Since emotions are strong motivators—either negatively or positively-- and saturate stories, points of view, decisions, and intentions, we turn to an exercise on emotions. If you are to listen to the feelings and emotions of clients, you should first be familiar with your own emotional states. A number of emotional states are listed below. You are asked to describe what you feel when you experience these emotions. Describe what you feel as _concretely_ as possible: How does your body react? What happens inside you? What do you feel like doing? Consider the following examples.

Example 1 — Accepted:

When I feel accepted,

· I feel warm inside.
· I feel safe.
· I feel free to be myself.
· I feel like sitting back and relaxing.
· I feel I can let my guard down.
· I feel like sharing myself.
· I feel some of my fears easing away.
· I feel at home.
· I feel at peace.
· I feel my loneliness melting away.

Example 2 — Scared:

When I feel scared,

· my mouth dries up.
· my bowels become loose.
· there are butterflies in my stomach.
· I feel like running away.
· I feel very uncomfortable.
· I feel the need to talk to someone.
· I turn in on myself.
· I'm unable to concentrate.
· I feel very vulnerable.
· I sometimes feel like crying.

1. Choose four of the emotions listed below, or feel free to use other emotions not on the list. Attempt to analyze the emotions you have difficulty with. It's important to listen to yourself when you are experiencing emotions that are not easy for you to handle.

42

2. Recall situations in which you have actually experienced each of these emotions.
3. Then, as in the example above, write down what you've felt when experiencing these emotions.

1. accepted	12. disappointed	23. lonely
2. affectionate	13. free	24. loving
3. afraid	14. frustrated	25. rejected
4. angry	15. guilty	26. respected
5. anxious	16. hopeful	27. sad
6. attracted	17. hurt	28. satisfied
7. bored	18. inferior	29. shy
8. competitive	19. interested	30. shocked
9. confused	20. intimate	31. superior
10. defensive	21. jealous	32. suspicious
11. desperate	22. joyful	33. trusting

The reason for this exercise is to sensitize you to the wide variety of ways in which clients express and name their feelings and emotions.

Although the feelings and emotions of clients (not to mention your own) are extremely important, sometimes helpers concentrate too much, or rather too exclusively, on them. Feelings and emotions need to be understood, both by helpers and by clients, in the *context* of the experiences and behaviors that give rise to them. On the other hand, when clients hide their feelings, both from themselves and from others, then it is necessary to listen carefully to cues indicating the existence of suppressed, ignored, or unmanaged emotion.

LISTENING THOUGHTFULLY TO CLIENTS' STORIES

Listening thoughtfully means identifying the key elements of clients' stories —experiences, behaviors, feelings —and the relationships among them. Thoughtful listening is a function of empathy.

EXERCISE 5-7: LISTENING FOR CORE MESSAGES

Core messages are the main points of a client's story. The ingredients of core messages are key experiences and key behaviors, coupled with the key feelings or emotions associated with those experiences and behaviors. In this exercise you are asked to "listen to" and identify the key experiences and behaviors that give rise to the client's main feelings.

1. Listen very carefully to what the client is saying.
2. Identify the client's key experiences, and what they say is happening to them.
3. Identify the client's key behaviors, what they are doing, not doing, or failing to do.
4. Identify the key feelings and emotions associated with these experiences and behaviors.

Example: A 27-year-old man is talking to a minister about a visit with his mother the previous day. He says, "I just don't know what got into me! She kept nagging me the way she always does, asking me why I don't visit her more often. As she went on, I got more and more angry. (He looks away from the counselor down toward the floor.) I finally began screaming at her. I told her to get off my case. (He puts his hands over his face.) I can't believe what I did. I called her a bitch. (Shaking his head.) I called her a bitch several times and then I left and slammed the door in her face."

Key experiences: Mother's nagging.

Key behaviors: Losing his temper, yelling at her, calling her a name, slamming the door in her face.

Feelings/emotions generated: He now feels embarrassed, guilty, ashamed, distraught, disappointed with himself, remorseful. (Note carefully: This man is not at this moment *expressing* anger. Rather he is talking *about* his anger, the way he lets his temper get away from him.)

Now do the same with the following cases.

1. A 40-year-old woman, married with no children, has had several sessions with a counselor. She went because she was bored and felt that all the color had gone out of her life. In a later session she says this: "These counseling sessions have really done me a great deal of good! I've worked hard in these sessions, and it's paid off. I enjoy my work more. I actually look forward to meeting new people. My husband and I are talking more seriously and decently to each other. At times he's even tender toward me the way he used to be. Now that I've begun to take charge of myself more and more, there's just so much more freedom in my life!"

Client's key experiences:

_____

_____

_____

Client's key behaviors:

_____

_____

_____

How does the client feel about these experiences and behaviors?

_____

_____

_____

2. A 20-year-old college student, who has volunteered to work with struggling high school students, finds that he is learning more than he imagined he would. He tells his college advisor what it's like: "These kids deal with problems I have never had to face. Some days we'll start by working on some bit of homework but before we're done they're talking about being afraid of the gangs on the way to and from school. This one kid starts and ends each day by helping his mom who has MS. In the morning he gets her up and dresses her and in the evening he gets her ready for bed. Some days I do as much listening as talking. I go home wondering about my own life. I get embarrassed when I think of the ways I whine about my own problems. Many of these kids have a type of maturity about them you don't see in people who have lived as comfortably as I have. Boy do I have a lot to learn!"

Client's key experiences:

_____

_____

_____

Client's key behaviors:

_____

_____

_____

How does the client feel about these experiences and behaviors?

_____

_____

_____

3. A 38-year-old woman, unmarried, talking about a situation with a friend: "My best friend has just turned her back on me. And I don't even know why! (said with great emphasis) From the way she acted, I think she has the idea that I've been talking behind her back. I simply have not! (also said with great emphasis) Damn! This neighborhood is full of spiteful gossips. She should know that. If she's been listening to those foul-mouthed jerks who just want to stir up trouble. . . . She could at least tell me what's going on. I'm tempted to tell to go jump in the lake. The nerve of her!"

Client's key experiences:

_____

_____

_____

Client's key behaviors:

_____

_____

_____

How does the client feel about these experiences and behaviors?

_____

_____

_____

4. A 64-year-old man, who has been told that he has terminal lung cancer, is speaking to a medical resident: "Why me? I'm not even that old! I keep looking for answers and there are none. I've sat for hours in church and I come away feeling empty. Why me? I don't smoke or anything like that. (He begins to cry.) Look at me. I thought I had some guts. I'm just a mess. Oh God, why terminal? What are these next months going to be like? (Pause, he stops crying.) Why would you care! I'm just a failure to you guys."

Client's key experiences:

_____

_____

_____

Client's key behaviors:

_____

_____

_____

How does the client feel about these experiences and behaviors?

_____

_____

_____

5. A 54-year-old woman, talking to a counselor about a situation at work: "I don't know where to turn. They're asking me to do things at work that I just don't think are right. If I don't do them—well, I'll probably be let go. And I don't know where I'm going to get another job at my age in this economy. But if I do what they want me to, I think I could get into trouble, I mean legal trouble. I'd be the fall guy. My head's spinning. I've never had to face anything this before. . . . Where do I turn?"

Client's key experiences:

_____

_____

_____

46

Client's key behaviors:

_____

_____

_____

How does the client feel about these experiences and behaviors?

_____

_____

_____

_____

EXERCISE 5-8: PROCESSING WHAT YOU HEAR—THOUGHTFUL LISTENING

Listening thoughtfully goes beyond being able to recall what the client has said. It means being able to discriminate between the details of the client's story and the really meaningful points—the core messages. This exercise is done with a learning partner.

1. Read the section in the book about "Processing What You Hear: The Thoughtful Search for Meaning." Pay particular attention to the extended example of Denise and Jennie.
2. Take the role of the client and, *through dialogue*, discuss a problem or unused opportunity from your own life with a learning partner. Spend between 5-10 minutes in the dialogue.
3. When you have finished, both you and your partner should jot down separately three essential elements of the story you have related—what each of you see as the core messages.
4. Compare what you have written, noting similarities and differences.
5. Next have your partner respond to the following questions:
    How well did I capture my partner's core messages? How accurate was I?
    What messages did I miss?
    What caused me to miss any given message?
6. Reverse roles and repeat the exercise.

LISTENING TO CLIENTS' POINTS OF VIEW, DECISIONS, AND INTENTIONS

Interweaved with clients' stories, goal-setting, and plans, are their points of view, decisions, and intentions. These need to be listened to carefully.

EXERCISE 5-9: LISTENING FOR REASONS, IMPLICATIONS, AND FEELINGS

1. A husband and wife have entered counseling to help them deal more effectively with their daughter. The daughter is fifteen and is struggling in school and becoming involved with older boys. At this point

47

the husband offers his point of view: "It seems to me that we both need to stay on top of her school work. We need to do a better job of that—no question. This business about her relationships with boys, though, I think should be handled primarily by you, her mother. I'll back up whatever needs to be said but there is a part of this which is best handled between two women. She might react negatively to you, but I think she'd be even more negative with me."

What is his point of view?

_____

_____

What are some possible reasons behind it?

_____

_____

_____

What are some possible implications of this point of view?

_____

_____

_____

2. A single parent talks to a school counselor about the fact that his son is being picked on by others in his class. He has discussed the facts of the matter and feels the counselor has a good grasp of what is going on. Not surprisingly he has a strong point of view as well. "If nothing is done, I'll have to take matters into my own hands and talk with the parents of the other boys. But that's not what I would prefer. I see this as a school matter and I will work with you and the teachers to see if this can be resolved in a sensible way. I know 'boys will be boys,' but this has gone too far already. I hate seeing him coming home so miserable."

What is the point of view?

_____

_____

What are the reasons behind it?

_____

_____

_____

What are some possible implications of this point of view?

_____

_____

_____

3. A manager of a project team has sought out the Director of Human Resources to discuss a matter of discipline. She uses the HR director to sound out a decision he is making. "I've decided I need to put Clarence on probation. I've wanted to avoid this but he's given me no choice. The performance improvement plan he's been on the last two months has been a bust. He still comes late or calls in sick with no notice. We're short-handed as it is and when he does this stuff, it nearly brings the operation to a halt. The others on his shift are up in arms and I've already said this is the next step. I don't want to do this, but I think I have to. Otherwise it would be unfair to everyone."

What is her decision?

_____

_____

What are the reasons behind it?

_____

_____

What are some possible implications of this decision?

_____

_____

_____

4. An 33-year-old unmarried woman has recently learned that she is HIV-positive. "When I first learned about this I was devastated. The end. The curtain is coming down. But one morning I woke up and said to myself, 'Do something.' I did. I got on the net and learned everything I could about this. I spent days studying it. And, you know, I'm going to beat it. I'm going to get physically fit. I'm going to get my head together. I'm going to take the drugs that keep you going. I'm not sure how I'm going to afford them, but I'll get them. I'm going to volunteer for some experimental approaches. I don't know how you get on the list, but somehow I will. I'm going to beat this thing."

What is she intending to do?

_____

_____

49

How is she going to make it work?

_____

_____

_____

How does she feel about all of this?

_____

_____

5. A 47-year-old man works in the construction trade and has been experiencing an increasing number of days when his knees and back ache to such a degree that he is hardly able to get dressed in the morning. He is talking with a friend and floats this possibility: "I don't know what other work I can get into at this point in my life, but I can't go on like this much longer. Ten years ago I'd have been made a supervisor and been able to hang in there but those days are gone. I'm taking painkillers every day and in the last year my right knee especially has been . . . well, I can't continue. What if I just made the move? I know I could do real estate appraisal. It might start slow but I think it would work."

What does he intend to do?

_____

_____

What are the reasons behind it?

_____

_____

_____

What are some of his concerns about what he intends to do?

_____

_____

LISTENING THOUGHTFULLY TO OPPORTUNITIES

Spotting and understanding clients' opportunities is just as important as understanding their problems. Often, exploring opportunities is the best way to manage problems.

EXERCISE 5-10: LISTENING TO KEY OPPORTUNITIES

The instructions for this exercise are the same as those for the previous exercise.

1. Ken, a 45-year-old man, has just been passed up for a promotion. He is talking with his supervisor, who was a strong supporter. He starts by talking about his disappointment. But his supervisor points out that staying in his present position will allow him to see a project he started through to completion. Ken shifts his focus to the opportunity side of the event: "You may have a point. In fact, during my entire career I've been so focused on my next job that I've tended to do a lot of switching horses in mid-stream—taking new jobs before projects were really completed. Don't get me wrong. I'm not giving up on seeking promotion around here. But, I like what I'm doing and now I can give it my undivided energy. I suppose that people have to see that I can see things through, that I can get results with the best of them. Moving too quickly in the past has prevented them from seeing me at my best."

What is the opportunity?

_____

_____

What are the upside and the downside to this situation?

_____

_____

How does Ken feel about it?

_____

_____

2. A 43-year-old woman has suffered a major financial setback in her small consulting business. She has just been offered a good position in a large consulting firm. "Lots of people would jump at this offer. The firm is solid, the job is challenging, and the pay is good. I'm really tempted. But I have really enjoyed being my own boss and building something out of nothing. If I take the offer, I'll have a boss again. Though I suppose that the economy has been my boss. The downturn is killing me. The little guys are the first to lose business. I do like the security I would have in the new job. But I know I won't be able to be as creative as I am when I'm on my own."

What is the opportunity?

_____

_____

What are the upside and the downside of the situation?

_____

_____

_____

How does she feel about it?

_____

_____

UNDERSTANDING CLIENTS' PROBLEMS AND OPPORTUNITIES THROUGH CONTEXT

EXERCISE 5-11: USING CONTEXT FOR THOUGHTFUL PROCESSING

Becoming a thoughtful processor requires two things — following the main points of the client's story, point of view, decision, or intention as it unfolds *and* at the same time, placing it in the context of their lives. Since we tend to know the context of our own lives, this exercise makes the point about thoughtful processing by asking you to focus on yourself.

1. Recall the main points if the section in the book about "Processing What You Hear: The Thoughtful Search for Meaning."
2. Share a problem or unexplored opportunity from your own life with a learning partner. Provide no background or context during the conversation.
3. Discuss the issue for 5-10 minutes.
4. When you have finished, tell your partner about the circumstances of your life that give added meaning to what you have shared. Discuss how the context changes the meaning of the problem or unexplored opportunity.
5. Reverse roles and give your learning partner a chance to do the same thing.

Example. A 26-year-old counseling trainee, who is a veteran from the conflict in Afghanistan, discusses how difficult it is to get back into the mainstream of society—family, community, culture, and now a professional education program. He talks about how a military mindset keeps overcoming him. For instance, many of the things he is now doing seem trivial in comparison to battlefield demands. He says, "In some significant ways I'm still in Afghanistan." He finds that engaging in leisure activities such as watching television or going to a movie with his girlfriend puts him on edge. He does not have the symptoms that go with PTSD—no "jitters, no nightmares," he says—but he just does not feel right.

Context. This counseling trainee comes from a Quaker background. He feels that some of his family members were disappointed when he joined the army and even more disappointed when he went to Afghanistan. But no one said anything to him. When he came home at the end of his first tour of duty, he was welcomed, certainly, but his community—family and friends—seemed different. He wondered, was *he* different, or did *they* feel differently about him? He joined the army, not because he was a great patriot, but because he felt that he had to do something for others "on the big screen." He felt guilty about being "a person of privilege" and wanted to help people who were suffering in ways he never dreamed of. But he did not share these sentiments with anyone. Keeping things to yourself had been part of the culture of his community.

Discuss with a partner how this context colors his story.

THE SHADOW SIDE OF LISTENING

Since effective helpers are active listeners, it helps to learn about your own listening style — both the upside and the downside.

EXERCISE 5-12: YOUR LISTENING AND PROCESSING STYLE AND ITS SHADOW SIDE

In this exercise you are asked to take an objective look at your style of listening, and the way you process what you hear.

1. What kind of listener are you? Give a brief, balanced description.

_____

_____

_____

_____

_____

_____

2. Re-read "The Shadow Side of Listening to Clients." Indicate below any areas of listening you think you should work on.

_____

_____

_____

_____

_____

3. With a learning partner, discuss your description of yourself as a listener and processor, together with your shadow-side self-assessment. Talk about the implications of your shadow-side traits. Indicate possible actions you can take to overcome whatever deficiencies you discover.

Chapter 6

EMPATHIC RESPONDING: WORKING AT MUTUAL UNDERSTANDING

The payoff of attending and listening lies in the helper's ability to communicate to clients an understanding of their stories, points of view, decisions, and intentions-- together with the feelings and emotions they generate. Responding with empathy to the key messages the client is trying to get across is one way of communicating your understanding. It is one important way in which you translate the *value* of empathy (discussed in Chapter 2) into behavior. Responding with empathy is a skill you'll need to develop as you work through the helping model. It is critical for establishing and developing relationships with clients, helping them clarify both problem situations and unexplored opportunities, setting goals, and developing and implementing strategies and plans.

EXERCISE 6-1: IDENTIFYING CORE MESSAGES IN CLIENTS' STORIES

In this exercise, you are asked to identify key experiences, behaviors, and/or feelings, and then translate them into a core message.

1. Listen carefully to the client's statement.
2. Identify key experiences, behaviors, and the feelings they generate.
3. Pull them together to create a statement of the client's core message.
4. After finishing, compare your responses with those of your learning partner. Then see if the two of you can improve upon your combined responses.

Example: Roula is recovering from a bad car accident in a rehabilitation facility. She is a week or two into a program that could easily last for a few months. In talking with one of the rehabilitation specialists, she says, "You told me it would be tough going. And I thought that I had prepared myself for it. I thought I had some courage in me. But now I can't find any grit at all. Doing even the smallest things takes so much effort! I keep breaking down and crying when I'm alone. I just keep giving in. . . . I'm just inches down a path that seems miles long. It seems endless."

> **Key experience(s):** finding the rehabilitation program so difficult
> **Key behavior(s):** failing to tap into the courage she thought she had; giving in to discouragement
> **Key feelings/emotions:** disappointment, discouragement, despair

Core message: Roula seems disappointed almost to the point of despair because she has failed to tap into the "stuff" within her that would keep her going during a very tough rehabilitation program.

Now do the same for the following cases.

1. This woman is about to go to her daughter's graduation from college. She is talking with one of the ministers of her church: "I never thought that my daughter would make it through. I've invested a lot of money in her education. It meant scrimping and saving and not having some of the things I wanted. But money is certainly not the issue. More to the point, I've put a lot of emotion into making this day happen. I had to do a lot of hand-holding to help her get through. There were times when neither of us thought this day would come. But the day has arrived!"

Key experience(s):

_____

_____

Key behavior(s):

_____

_____

Key feelings/emotions:

_____

_____

Core message:

_____

_____

2. An older man has just had his wallet stolen. He became disoriented and was taken to a hospital. He's talking with a social worker. "I had just cashed my paycheck and the money was in the wallet. I've had a streak of bad luck. My sister was in an auto accident last week. And I was on my way to visit my nephew. Ironically, he was detained by the police for a minor theft earlier this week. Now he's probably thinking I've given up on him, when I haven't. There has not been much good news at all for a while."

Key experience(s):

_____

_____

Key behavior(s):

_____

_____

Key feelings/emotions:

_____

_____

Core message:

_____

_____

_____

3. This woman is waiting for the results of some medical tests. She is talking to a hospital volunteer: "I've been losing weight for about two months and feeling tired and listless all the time. I'm afraid of what these tests are going to say. I know I've been putting them off. I just hate hospitals and this kind of stuff. Well, now the waiting is getting to me. I . . . well, I just don't know where I stand. Nobody said anything to me during the tests. I don't think that's a good sign. It's all so impersonal anyway."

Key experience(s):

_____

_____

Key behavior(s):

_____

_____

Key feelings/emotions:

_____

_____

Core message:

_____

_____

_____

4. A prospective employer has just found out that an applicant has a criminal record. The applicant is denied the position: "I had hoped that I would get the job and prove myself before anyone found out about my record. I guess I was just stupid. I've just received a call from him (the employer) telling me that I'm no longer being considered for the job. Well, I did what I thought was right. I never had the intention of deceiving anyone. I mean that I didn't think that I was doing anything wrong. . . . Well, even though I'm branded because of my record, I'm going to make it, somehow. I have to."

Key experience(s):

_____

_____

Key behavior(s):

_____

_____

Key feelings/emotions:

_____

_____

Core message:

_____

_____

5. A woman has just lost a custody battle for her only son. During the interview with a counselor, she seems almost in a daze: "I never dreamed that the court would award custody to my husband. I may not be a perfect mother, but I've done all the work in raising him so far. My husband has done nothing but give me a hard time over the years. He's so selfish and spiteful. I've been racking my brains trying to come up with something I can do. . . . Now he's won. . . . It's all over."

Key experience(s):

_____

_____

Key behavior(s):

_____

_____

Key feelings/emotions:

_____

_____

Core message:

_____

_____

_____

EXERCISE 6-2: USING A FORMULA TO COMMUNICATE EMPATHY

Empathy focuses on the client's core messages—key experiences and/or key behaviors, key points of view, key decisions, key intentions—plus the feelings and emotions they generate. In this exercise you are asked to use the formula explained in the text: "You feel . . ." (followed by the right emotion and some indication of its intensity) "because . . ." (followed by the key experiences and/or behaviors that give rise to the emotion).

1. Identify for yourself the speaker's key experiences, behaviors, points of view, decisions, intentions, and the feelings associated with them.
2. Formulate an empathic response. Even though you are writing the response, picture yourself actually talking to the person.

Example: A woman in a self-help group is talking about a relationship she had with a man. She says: "About a couple of months ago, he began being abusive, calling me names, describing my defects. To tell you the truth, that's why I joined this group, but I haven't had the courage to talk about it until now. The couple times I've tried to stand up for myself, he became even more abusive. He hasn't hit me or anything, but, anyway. So I've been just taking it, just sitting there taking it . . . like a dog or something. Do you think that this is just his bizarre way of getting rid of me? Why doesn't he just tell me?"

> **Key experience(s)**: being abused by companion, escalation in abuse
> **Key behavior(s)**: trying to stand up for herself, becoming passive, trying to figure out if this is his way of getting rid of her
> **Key feelings/emotions**: distraught, confused, angry

Empathic response (using formula): You feel angry and confused because the abuse came from out of the blue, and now you're wondering whether this is just his strange way of ending the relationship.

1. A first-year college student is talking to a counselor in the Center for Student Services. She has been talking about some of the difficulties of adjusting to college life. She comes from a small town and is attending a large state university. She speaks openly and seems to be in good spirits; she even smiles at times: "My friends seem so much more sophisticated than I am. Just the other night Mairead and Johanna showed up at the party in dresses *and* with dates. And there I was, alone and in jeans! I mean, just typical of me! . . . I guess I was more amazed than embarrassed. And I did have a good time anyway. It's like I have to learn a whole new set of rules. But nobody just comes out and says what they are."

Key experience(s):

_____

_____

Key behavior(s):

_____

_____

Key feelings/emotions:

_____

_____

Empathic response (formula):

_____

_____

2. A 66-year-old man is talking to a mental health counselor: "My wife died last year, and this year my youngest son went away to college. The other children are married. So now that I'm retired, I spend a lot of time rambling around a house that's . . . . (pauses, looks out the window for a while) . . . really too big for me. . . . You know, when I was working, there was a certain fullness to life. I always knew what to do. I, well . . . I made a difference. Now that I've got a comfortable retirement, I . . . ."

Key experience(s):

_____

_____

Key behavior(s):

_____

_____

Key feelings/emotions:

_____

_____

Empathic response (formula)

_____

_____

_____

3. A 33-year-old woman has been working with a counselor to find ways of handling sexual discrimination at work. The company seems to preach one philosophy but implement another. Up to this point she has had some modest success, but now she wants to discuss a setback: "It seems they let me push as long as I stayed with the little stuff and did it quietly. But last week I brought up discrimination in our monthly team meeting . . . and my boss clammed up. And he's not one of the worst offenders. He's been quite cool since then and always seems to be busy. All of a sudden there's quite a different atmosphere in the office."

Key experience(s):

_____

_____

Key behavior(s):

_____

_____

Key feelings/emotions:

_____

_____

Empathic response (formula):

_____

_____

_____

4. A 25-year-old woman is talking to an older confidante about her current boyfriend: "I can't quite figure him out. I still can't tell if he really cares about me, or if it's just about sex. It leaves everything up in the air. He's pleasant enough, but I'm not finding the substance that I thought was there. I do try to talk about serious things, things that interest me like what's happening in the world. The next thing I know we're back into trivia. I've been tempted to talk about this directly, but something holds me back."

Key experience(s):

_____

_____

Key behavior(s):

_____

_____

Key feelings/emotions:

_____

_____

Empathic response (formula):

_____

_____

_____

6. A young man has just been abandoned by his wife after only a year of marriage: "We've been married for about a year. . . . She left a note saying that this has not been working out. Just like that. You know, I thought that things were going fairly well. Not perfect, of course. We had our ups and downs, but everyone does. Maybe I was so busy at work that there were things I didn't notice. I don't know if there's someone else. I really don't know why. . . . I must really sound stupid. I have no idea what to do to get her back. You can't make someone love you."

Key experience(s):

_____

_____

Key behavior(s):

_____

_____

Key feelings/emotions:

_____

_____

Empathic response (formula):

_____

_____

_____

EXERCISE 6-3: USING YOUR OWN WORDS IN EMPATHIC RESPONSES

In this exercise you are asked to do two things:

1. First use the "you feel . . . because . . ." formula to communicate understanding to the client.
2. Then recast your response in your own words while still identifying both core messages.

Example: A 31-year-old married woman is having a "solo" meeting with a marriage counselor, to review how things have been going from her point of view. At one point she says: "I can't believe it! You know when Tom and I were here last week we made a contract that said that he would be home for supper every evening and on time. Well, he came home on time every day this past week. I never dreamed that he would live up to his part of the bargain so completely! I've even started making better meals!"

> **Formula.** "You feel great because he really stuck to the contract!"
> **Non-formula.** "Making good on his word has given you quite a boost!"

Now imagine yourself listening intently to each of the clients quoted below. First use the "You feel . . . because . . ." formula; then use your own words. Try to make the second response sound as natural (as much like yourself) as possible. After you use your own words, check to see if you have both a "you feel" part and a "because" part in your response.

1. A 28-year-old woman is talking with a colleague about her job: "It's not a big thing. But this is the third time this month I've been asked to change hours with her. I said yes, of course. But it certainly seems to indicate who is more important there. Why does it always have to be me who defers to her?"

Use the formula.

_____

_____

Use your own words.

_____

_____

2. A man, who has been suffering from depression for a number of months, is talking with a psychologist: "I've seen my doctor a couple of times. But now since the depression isn't getting any better, he wants to give me this drug. . . . But I don't even take aspirin if I don't have to. Somehow drugs and I don't mix. If I had pneumonia or some infection, it would be one thing. But this is all in my head. I don't want to be seen as a psycho!"

Use the formula.

_____

_____

Use your own words.

_____

_____

3. A counseling trainee is talking to her instructor. The training group has just been introduced to the skill of challenging clients. They have practiced self-challenge and have just begun using challenge with one another: "For the first time in this program I having trouble and it's bothering me. What I've found out is that I have no problem with self-challenge. In some ways I have been doing that for years. But now I see that maybe I've been picking on myself, you know, being too much of a perfectionist. And I'm finding it really hard to challenge the members of my group. Oh, I know the theory. But doing it is hard. It's almost as if I feel like they won't like me if I do it well."

Use the formula.

_____

_____

Use your own words.

_____

_____

4. A fellow counseling trainee has been talking with you about the counselor training program. He calls you up and says he wants to talk with you right away. When you meet, he says: "You know, tomorrow we're going to start talking about our own problems. Well, the kinds of things that could stand in the way of being good helpers. . . . Well, there are things I'd rather not say in the group. . . . I even had to screw up my courage to talk with you. . . . You know, a couple of issues I'd rather not talk about *could* stand in the way of my being the kind of counselor I should be. . . . But I'm just not ready. I could be dishonest."

Use the formula.

_____

_____

Use your own words.

_____

_____

5. A man, Alfred, is being forced out of the company where he has held a senior position for almost twenty years: "It came out of nowhere. I came in two days ago, and there on my desk was a notice from the venture capital group that bought us, saying that I was terminated. It was as impersonal as that. . . . I kept reading it over and over again like I was in some dream. I tried to call my boss but couldn't get her. Later I found out that she had been let go, too. The notice said I get some outplacement help. What kind of word is that! It's a fancy word for dumping."

Use the formula.

_____

_____

Use your own words.

_____

_____

EXERCISE 6-4: EMPATHIC RESPONSES WITH CLIENTS FACING DILEMMAS

Clients often talk about complicated issues. When this is the case, it is essential to listen even more carefully, in order to pick up and respond to core messages. For example, clients often talk about conflicting values, complicated experiences, behaviors, points of view, decisions, intentions, and complex emotions. Responding with empathy means communicating an understanding of the conflict.

1. Identify the conflict or dilemma.
2. Use the empathic response formula to communicate understanding.
3. Express the response in your own words.

Example: A 32-year-old woman is talking to a counselor about adopting a child: "I'm going back and forth, back and forth. I say to myself, 'I really want a child,' but then I think about Bill [her husband] and his reluctance. He so wants our own child, and is so reluctant to raise someone else's. We don't even know why we can't have children. But the fertility specialists don't offer us much hope. At times when I so want to be a mother I think I should marry someone who would be willing to adopt a child. But I love Bill and don't want to point an accusing finger at him."

> **The conflict or dilemma.** She believes that she runs the risk of alienating her husband if she insists on adopting a child, even though she strongly favors adoption.
> **Formula**. "You feel trapped between your desire to be a mother and your love for your husband."
> **You own words.** "You're caught in the middle. Adopting a child would solve one problem but perhaps create another."

1. A 30-year-old factory worker is talking to a counselor about his job: "Work is okay. I do make a good living, and both my family and I like the money. My wife and I are both from poor homes, and we're living much better than we did when we were growing up. But the work I do is the same thing day after day. I may not be the world's brightest person, but there's a lot more to me than I use on those machines."

The conflict.

_____

_____

The formula.

_____

_____

Your own words.

_____

_____

_____

2. A 54-year-old patient at a mental hospital, who has spent five years in the hospital; is talking to the members of an ongoing therapy group. Some of the members have been asking him what he's doing to get out. He says, "I don't know why you're trying to push me out of here. To tell the truth, I like it here. I'm safe and secure. So why are so many people here so damn eager to see me out? . . . Is it a crime because I feel comfortable here? (Pause, then in a more conciliatory voice.) I know you're all interested in me. I see that you care. But do I have to please you by doing something I don't want to do?"

The conflict.

_____

_____

The formula.

_____

_____

Your own words.

_____

_____

_____

3. A juvenile probation officer is speaking to a colleague: "These kids drive me up the wall. Sometimes I think I'm really stupid for doing this kind of work. They taunt me. They push me as far as they can. To some of them I'm just another 'pig.' But every time I think of quitting—and this gets me—I know I'd

miss the work and even miss the kids one way or another. When I wake up in the morning, I know the day's going to be full and it's going to demand everything I've got."

The conflict.

_____

_____

The formula.

_____

_____

Your own words.

_____

_____

_____

EXERCISE 6-5: RESPONDING WITH EMPATHY IN EVERYDAY LIFE

If communicating empathy is to become a part of your natural communication style, you will have to practice it outside formal training sessions. It must become part of your everyday communication style, or it will tend to lack candor in helping situations. Practicing empathic communication "out there" is a relatively simple process.

1. **The improbability of empathic responding.** Responding to others with empathy is not a normative response in everyday conversations, even among people who otherwise put into practice the value of empathy. Find this out for yourself. Observe everyday conversations. Count how many times empathy is used as a response in any given conversation.
2. **Your own use of empathy.** Next, try to observe how often you yourself use empathy as part of your normal style. In the beginning, don't necessarily try to increase the number of times you do so. Merely observe your usual behavior. What part does empathic responding normally play in your interpersonal communication style?
3. **Increasing your empathic responses.** Begin to increase the number of times you respond to others with empathy. Be as natural as possible. Do not overwhelm others with this response; rather try to incorporate it gradually into your style. You will probably discover that there are quite a few opportunities for responding with empathy without being phony. Find some way of keeping track of your progress.
4. **The impact of empathy.** Observe the impact your use of empathy has on others. Don't use others for the purpose of experimentation, but, as you gradually increase your use of this communication skill naturally, notice how it influences and enriches your conversations. What impact does it have on you? What impact does it have on others?
5. **Learnings.** In a forum set up by the instructor, discuss with your fellow trainees what you are learning from this experiment.

If responding with empathy becomes part of your communication style "out there," then you should appear more and more natural in using empathy in the training program, both in playing the role of the helper and in listening and providing feedback to your fellow trainees. On the other hand, if you use empathy only in the training sessions, it will most likely remain artificial.

Chapter 7

THE ART OF PROBING AND SUMMARIZING

Problems and opportunities are better managed when they are clear and specific. If a husband and wife want to deal with the poor communication they have with each other in their marriage, then the key issues relating to communication need to be spelled out in terms of specific experiences, behaviors, points of view, decisions, intentions, and feelings in specific situations. Clients need to focus on concrete-- rather than abstract-- issues. Counselors use probing remarks and questions to help clients "fill in the picture," turning vague statements into clear and specific ones. Vaguely described problems will only lead to vague solutions. Read Chapter 7 in *The Skilled Helper* before doing these exercises. This section has three parts: problem and opportunity clarity; probing; and summarizing.

PROBLEM AND OPPORTUNITY CLARITY

EXERCISE 7-1: SPEAKING CONCRETELY ABOUT PROBLEMS

This exercise is designed to help you recognize the difference between vagueness and concreteness in your clients' statements. In this exercise you are asked to focus on your own story in terms of your own experiences, behaviors, points of view, decisions, intentions, and feelings.

1. Give two vague statements of a problem situation.
2. Turn each vague statement into a concrete and specific statement, in terms of key experiences, behaviors, points of view, decisions, intentions, and feelings.
3. Focus on issues that can make a difference in your life.

Example 1: Karen writes about her need to control situations.

> **Vague statement of problem situation**: "I tend to be a bit domineering at times."
> **Concrete statement**: "I try, usually in subtle ways, to get others to do what I want to do. I even pride myself on this. In conversations, I take the lead. I interrupt others, jokingly and in a good-natured way, but I make my points. If a friend is talking about something serious when I'm not in the mood to hear it, I change the subject."

Example 2: Jamie talking about how the training group affects him.

> **Vague statement of problem situation**: "I get anxious when I give feedback in training groups."
> **Concrete statement**: "I feel hesitant and embarrassed whenever I want to give feedback to other trainees, especially if it is in any way negative. When the time comes, my heart beats faster and my palms sweat. I feel like everyone is staring at me, though I know they're not. I feel under pressure in the group. Outside I have a live-and-let-live mentality. I'm not used to giving feedback to anyone."

In the spaces below, explore two issues that are related to some problem situation(s) of your own. Try to choose issues that might affect the quality of your helping.

1. Vague statement.

_____

_____

Concrete statement.

_____

_____

_____

_____

_____

_____

2. Vague statement.

_____

_____

Concrete statement.

_____

_____

_____

_____

_____

Share one or two of these with a learning partner. Get feedback on how clear your second statement is. If you do not think that your partner's statement is as clear as it could be, use probing remarks and questions to help them make their statement more clear.

EXERCISE 7-2: SPEAKING CONCRETELY ABOUT UNUSED OPPORTUNITIES

In this exercise, you are asked to speak about some of your unused/ underused resources or opportunities. As in the exercise above, start with a vague statement, and then clarify it with the kind of detail needed to serve the opportunity-development process. Try to choose unused resources or opportunities that might add value to your role as helper.

1. Give a vague statement of some unused/ underused resource or opportunity.
2. Turn the vague statement into a concrete, specific statement in terms of key experiences, behaviors, points of view, decisions, intentions, and feelings.

Example 1: Jane, a counselor trainee, discusses her communication style.

> **Vague statement of unused opportunity**: "There is a very good communicator inside me that doesn't get out enough."
>
> **Concrete statement**: "This course has shown me that I am already fairly good at many of the communication skills we are learning. I listen well. I'm empathic. From what I understand about challenge, I think that I can even challenge people without turning them off. But so often I choose to be passive. It groups I'm often an onlooker rather than a participant. I found out that I can't get away with that in this group. I think I need to become much more assertive in order to be an effective learner and helper."

Example 2: Austin, in his fourth year of a doctoral program in clinical psychology, has also taken a couple of courses in organizational psychology. He has just finished his internship.

> **Vague statement of possible opportunity**: "One thing I learned during my internship rotations is that mental-health organizations are not what they could be. But I'm not sure I can do anything about it without violating my commitment to people."
>
> **Concrete statement**: "The organizational psychology courses I took made it clear to me that some mental-health centers are poorly-run organizations. Even with a little background I knew some things I could do to help them be more efficient and effective. In fact, the whole area of applying effective management and organizational ideas to mental-health systems appeals to me enormously. But I feel guilty about pursuing that possibility after graduation. It's like betraying the reasons for going into clinical psych in the first place — mainly to help others. I really liked dealing with individuals and groups during my internship, but I was distracted by how poorly these places are administered. A number of psychologists I met thought that administration was for the birds, you know, for lesser people."

In the spaces below, deal with two instances of unused/ underused resources or opportunities. Choose opportunities that could add value to your role as helper.

1. Vague statement.

_____

_____

Concrete statement.

_____

_____

_____

_____

_____

_____

2. Vague statement.

_____

_____

Concrete statement.

_____

_____

_____

_____

_____

Share one or both of these with your learning partner. Get feedback on how clear your second statement is. If you do not think that someone's statement is as clear as it could be, use probing remarks and questions to help your fellow trainee make his or her statement clearer.

PROBES

EXERCISE 7-3: PROBING FOR KEY ISSUES AND CLARITY

A probe is a statement or a question that invites a client to discuss an issue more fully. In the previous exercises you were asked to exercise this skill on yourself. Probes are ways of getting at important details that clients do not think of or are reluctant to talk about. They can be used at any point in the helping process to clarify and explore issues, search for missing data, expand perspectives, and point toward possible client actions. However, an overuse of probes can lead to gathering a great deal of irrelevant information. The purpose of a probe is not data for its own sake, but information—experiences, behaviors, points of view, decisions, intentions, and feelings—that serve the process of problem management and opportunity development.

In this exercise, brief problem situations, together with a bit of context, will be presented. Your job is to formulate possible probes to help clients move forward in the helping process.

1. First use empathic responses to understand the client.
2. Formulate a probe that might help the client further identify or explore key issues, experiences, behaviors, points of view, decisions, intentions, and feelings.
3. Briefly state how your probe might help the client move forward.

Example: A 24-year-old man complains that he is severely tempted to continue experimenting sexually with women other than his wife: "Although I have not had an extended affair, I have had a few sexual encounters and feel that some day I might have an affair. I don't blame myself or my wife, though I'm not sure I'll ever find the kind of sexual satisfaction I want with her. I've never had such strong sexual urges. I thought that once I was married, the sexual thing would be taken care of. That's certainly not the way things have turned out. I'm not sure where I'm headed."

> **Empathic statement**: "It sounds like your sexual urges are so strong that right now that they're in the driver's seat. But you're puzzled and are not quite sure what all this is means and where it is going."
> **Possible probe**: "What do you think are some of the possible consequences of the course you've taken?"
> **How might it help the client move forward**? To help the client begin thinking through some of the implications of the decisions he's made about his sexual life.

1. Grace, a 19-year-old, unmarried, first-year college student, comes to counseling because of an unexpected and unwanted pregnancy: "Right now I realize that the father could be one of two guys. That probably says something about me right there. I'm not sure what I want to do about the baby. I haven't told my parents yet, but I think that they will be very upset, but in the end sympathetic. But I've gotten myself into this mess and I have to get myself out."

Empathic statement.

_____

_____

Possible probe.

_____

_____

How might it help the client move forward?

_____

_____

2. You are a counselor in a halfway house. You are dealing with Tom, a 44-year-old man, who has just been released from prison where he served two years for armed robbery. This was his first encounter with the law. He has been living at the halfway house for two weeks. The halfway house experience is designed to help him reintegrate himself into society, and living in the house is voluntary. The immediate problem is that Tom came in drunk a couple of nights ago when he was supposed to be job-hunting, and drinking is against the rules of the house. When you talk to him, he says "I know this looks bad. But I don't think it's as bad as it looks. There was no liquor inside prison. This was my first encounter with

alcohol since I got out. I guess I just went overboard. I shouldn't have done it, but it's not the end of the world. Or it shouldn't be."

Empathic statement.

_____

_____

Possible probe.

_____

_____

How might it help the client move forward?

_____

_____

3. Arnie is a born-again Christian. He has begun to do a fair amount of informal preaching at his place of employment. While some of his co-workers sympathize with his views, others are turned off. Since he feels that he is being driven by a "clear vision," he becomes more and more militant. His supervisor has cautioned him a couple of times, but this has done little to change Arnie's behavior. Finally, he is given an ultimatum to either talk to one of the counselor's in the Employee Assistance Program about these issues, or be suspended from his job. He says to the counselor, "I have a duty to spread the word. And if I have a duty to do so, then I also have the right. I'm a good worker. In fact, I believe in hard work. So it's not like I'm taking time off for the Lord's work. Now what's wrong with that?"

Empathic statement.

_____

_____

Possible probe.

_____

_____

How might it help the client move forward?

_____

_____

4. A trainee in the counselor training program makes this statement in his small group: "I do not take criticism well. When I receive almost any kind of negative feedback, I usually smile and seem to shrug it off, but inside I begin to pout. Also, deep inside, I put the person who gave me the feedback on a 'list.' Those on the list have to pay for what they did. I find this hard to admit, even to myself. It sounds so petty. For instance, two weeks ago in the training group I received some negative feedback from Cindy. I

felt angry and hurt because I thought she was my friend. Since then I've tried to see what mistakes she makes here. I've even felt a bit disappointed because I haven't been able to catch her in any kind of glaring mistake. It goes without saying that I'm embarrassed to say all this."

Empathic statement.

_____

_____

Possible probe.

_____

_____

How might it help the client move forward?

_____

_____

5. Renata is talking with a counselor about her relationship with her mother: "I feel guilty and depressed whenever my mother calls and implies that she's lonely. I then get angry with myself for giving in to guilt so easily. Then the whole day has a cloud over it. I get nervous and irritable and take it out on others. Or I brood. Sometimes it even interferes with my work. Then I do what I have to."

Empathic statement.

_____

_____

Possible probe.

_____

_____

How might it help the client move forward?

_____

_____

EXERCISE 7-4: EMPATHIC RESPONSES AND TWO KINDS OF PROBES

This exercise asks you to combine several skills: the ability to be share understanding through empathic responses; to identify areas needing clarification; and to use both question-type and statement-type probes to help clients clarify their stories, points of view, decisions, and intentions, together with how they feel about all of these.

1. First reply to the client with empathy.
2. Use a question-type probe in open-ended question form to help the client explore or clarify some issue.
3. Convert the question-type probe into a statement-type probe.
4. Indicate how the probe might help the client move forward.

Example: A 25-year-old law student is talking to a school counselor: "I learned yesterday that I've flunked out of school and that there's no recourse. I've seen everybody, but the door is shut tight. What a mess! I know I haven't gotten down to business the way I should. This is my first year in a large city and there are so many distractions. And school is so competitive. I have no idea how I'll face my parents. They've paid for my college education and this year of law school. I'll have to tell them that it's all down the drain."

> **Empathic response**: "In your eyes the punishment doesn't seem to fit the crime. And it all sounds, not just desperate, but so final."
> **Question probe**: "Which people did you see and which doors were shut?"
> **Statement probe**: "I'm not sure who you mean by 'everybody' and it's not clear which doors are completely shut."
> **How might the probe help the client move forward?** It might help the client identify possibilities that he has overlooked because of the emotionality surrounding the situation.

1. A high school senior is speaking to a school counselor: "My dad told me the other night that I looked relaxed. Well, that's a joke. I don't feel relaxed. There's a lull right now, because of semester break, but next semester I'm signed up for two math courses, and math really rips me up. But I need it for science since I want to go into pre-med."

Empathic response.

_____

_____

Question probe.

_____

_____

Statement probe.

_____

_____

How might it help the client move forward?

_____

_____

2. A 27-year-old woman is talking to a counselor about a relationship that has just ended (she speaks in a rather matter-of-fact voice): "About three weeks ago I came back from visiting my parents who live in Nevada and found a letter from my boyfriend Gary. He said that he still loves me but that I'm just not the person for him. In the letter he thanked me for all the good times we had together these last three years. He asked me not to try to contact him because this would only make it more difficult for both of us. End of story. I guess I've let my world collapse. People at work have begun complaining about me. And I've been like a zombie most of the time."

Empathic response.

_____

_____

Question probe.

_____

_____

Statement probe.

_____

_____

How might it help the client move forward?

_____

_____

3. A 25-year-old married man is talking to a counselor about trouble with his mother-in-law: "The way I see it, she is really trying to destroy our marriage. She's so conniving. And she's very clever. It's hard to catch her in what she's doing. You know, it's rather subtle. Well, I've had it! If she's trying to destroy our marriage, she's getting pretty close to achieving her goal."

Empathic response.

_____

_____

Question probe.

_____

_____

Statement probe.

_____

_____

How might it help the client move forward?

_____

_____

4. A 31-year-old woman is talking to her friend, an older woman: "I just can't stand my job any more! My boss is so unreasonable. He makes all sorts of silly demands on me. The other women in the office are so stuffy, you can't even talk to them. The men are either very blah or after you all the time, you know, on the make. The pay is good, but I don't think it makes up for all the rest. It's been going on like this for almost two years."

Empathic response.

_____

_____

Question probe.

_____

_____

Statement probe.

_____

_____

How might it help the client move forward?

_____

_____

5. A 45-year-old man, who has lost his wife and home in a tornado, has been talking about his loss to a social worker: "This happened to a friend of mine in Kansas about ten years ago. He never recovered from it. His life just disintegrated and nobody could do anything about it. . . . It was like the end of the world for him. You never think it's going to happen to you."

Empathic response.

_____

_____

Question probe.

_____

_____

Statement probe.

_____

_____

How might it help the client move forward?

_____

_____

EXERCISE 7-5: PROBING FOR ACTION

Remember that each step of the helping process should be a kind of stimulus for client action-- the "little" actions that precede formal action, based on a plan. In this exercise, you are asked to use both empathic responses and probes to help clients consider actions they might take to move forward. Probes are very useful in helping clients develop a solution-focused approach in the helping process.

Example. A 44-year-old divorced woman is talking to a counselor about her drinking. This is the second session. She has spent a lot of time telling her story. She has already admitted that she engages in evasions and sometimes simply lies about her problems. She says, "Actually, it's a relief to tell someone the truth. I don't have to give you any excuses or make the story sound right. I drink because I like to drink; I'm just crazy about the stuff, that's all. But I'm under no delusions that telling you is going to solve anything. When I get out of here, I know I'm going straight to a bar and drink. Some new bar, new faces, some place they don't know me."

> **Empathic response.** "So, for you, drinking is a way of life. No use making excuses. That's the way it is. So you're headed to a bar."
> **Probe for action**. "If a friend had heard you talking and challenged you by saying, 'Hey! You're better than this. Don't go to a bar. Do something constructive, something creative for a change," what could you do instead of going to the bar?" What might that be?"

1. A 57-year-old man is talking to a counselor about a family problem: "My younger brother, he's 53, has always been a kind of bum. He's always poaching off the rest of the family. Last week my unmarried sister told me that she'd given him some money for a 'business deal.' Business deal, my foot! I'd like to get hold of him and give him a good kick! Oh, he's not a vicious guy. Just weak. He's never been able to get a fix on life. But he's got the whole family in turmoil now, and we can't keep going through hell for him."

Empathic response.

_____

_____

Probe for action.

_____

_____

2. A 25-year-old graduate student is speaking to an instructor whom he trusts: "I have two term papers due tomorrow. I'm giving a report in class this afternoon. My wife is down with the flu. And now I find out that the program director wants to talk with me about my progress in the program. I've been working a couple of part-time jobs—I've got to just to keep things going. I know I'm behind, but I still think I can handle things. I hope they're not going to give me any kind of ultimatum."

Empathic response.

_____

_____

Probe for action.

_____

_____

3. A 49-year-old man is talking to a rehabilitation counselor after an operation that has left him with one lung: "I'll never be as active as I used to be. But at least I'm beginning to see that life is still worth living. I have to take a long look at the possibilities, no matter how much they've narrowed. I can't explain it, but there's something good stirring in me."

Empathic response.

_____

_____

Probe for action.

_____

_____

4. Mark and Lisa, a married couple, both 33 years old, have spent years attempting to have children and have finally adopted a baby girl, Andrea. Their relationship, which up to then seemed quite good, has begun to disintegrate. Mark has made some cracks about "the stranger in the house." Andrea has proved to be a somewhat difficult baby. Lisa feels exhausted and blames Mark for not helping her. They are both thinking about divorce now, but feel very guilty because of the child. Both of them say, "If only we had never adopted Andrea."

Empathic response.

_____

_____

Probe for action.

_____

_____

EXERCISE 7-6: USING PROBES TO HELP OTHERS COMMUNICATE BETTER

This is an exercise that can be done both in the training group and in your everyday life. You most likely run into people who are not very good at dialogue. You can use probes to help them do better.

1. Find a person who is not a good communicator, someone not good at dialogue.
2. Use both empathic responses and probes to help this person engage in the four requirements for true dialogue: engage in turn taking, connect with what you have to say, stay open to mutual influence, and, in the end, co-create some kind of conversational outcome with you.
3. After you have engaged in a couple of these conversations, discuss your successes and your failures with a learning partner. What do you have to do to become better at helping less-communicative people engage in dialogue? Ultimately, this means helping clients engage in dialogue as they tell their stories, reveal their points of view, discuss their decisions, and reveal their intentions.

Example. Trudy, a counselor trainee, is talking with Nathan, a friend of hers in the neighborhood. Nathan, who always has very strong opinions, has been talking almost nonstop about why he thinks that the government is going to make a big mistake by over-regulating the financial industry, in a response to the financial debacle. Trudy uses the skills she is learning to try to help Nathan have a dialogue with her.

> **Nathan** (continuing at a fast pace): "Take compensation. Passing laws about how much people should be paid is not the function of government. Why not . . .."

> **Trudy** (interrupting): "Nathan, your main point seems to be that government intervention just doesn't work. The times they've tried to run the banks, they've done a poor job. So that's what

they shouldn't do. What do you think that government agencies should do? What would you do to make the financial industry more responsible?"

Nathan: "I'm not sure that I'd do anything. I mean, I wish that the system would just straighten itself out. There are a lot of very smart people in finance."

Trudy: "I don't doubt that, but if they are so smart, why did they let things get out of hand? A lot of the bankers have been admitting that mistakes have happened, even though they never seem to take person responsibility for any of them."

Nathan: "OK, mistakes—a lot of mistakes—were made, but why should big government make things worse by making their own mistakes? We'd lose our financial leadership position. Let me tell you how it works. . ."

Trudy (interrupting): "Who says that we have to be financial leaders? What about ethics? What about the common good? What about responsibility? I'm not sure what you mean by leadership."

Nathan: "Leadership means the best. The most powerful. The most influential."

Trudy: "It seems that the financial industry "influenced" a lot of people to get mortgages they could not afford."
Nathan: "All right. I'm not trying to say that the financial industry is perfect. There are a lot of warts. What we need is some kind of balanced approach to reform. And I don't see that happening right now. Government is overwhelming the industry."

Trudy: "So 'balance' is the important operating term. The question is, how do we get the balance right? A lot of different voices have to contribute to the solution."

Nathan: "Yes, but we need to balance the power of the voices. We need focus. Don't get me wrong. I think that the common good should play an important role here, but we need to define what we mean by the common good."

Trudy: "Yes, it sounds like we need to discuss a lot of things we take for granted. The common good is just one of them. Justice, fairness, equity, community—there's a lot of things we need to define if we are going to negotiate sensibly."

Nathan: "There's a good word. Negotiate. Some of the ads about negotiation-skills programs seem to define negotiating as winning. That's not right."

Trudy: "Yes, if it means winning, then you need to do whatever it takes to win. That's one of the things that got is into this mess in the first place."

Nathan: "It seems that we need to take the whole financial system apart conceptually and then put it back together again. Lack of transparency and mutual understanding has caused a lot of trouble."

And on they go. Discuss Trudy's approach here with your learning partner. What did she do right? What did she do wrong? What could she have done to improve the quality of the dialogue?

SUMMARIZING

At a number of points throughout the helping process, it is useful for helpers to summarize-- or to have clients summarize-- the principal points of their interactions. This places clients under pressure to focus and move on. Since summarizing can place pressure on clients to move forward, it is sometimes a type of challenge.

EXERCISE 7-7: SUMMARIZING AS A WAY OF PROVIDING FOCUS

This exercise assumes that trainees have been using the skills of sharing empathic responses and probing in order to help one another tell their stories, share their points of view, discuss their decisions, and spell out their intentions.

1. The total training group is divided into subgroups of three.
2. There are three roles in each subgroup: helper, client, and observer.
3. The helper spends about eight minutes counseling the client. The client should continue to explore one of the problem areas he or she has chosen to deal with in the training group.
4. At the end of four minutes, the helper summarizes the principal points of the interaction. Helpers should try to make the summary both accurate and concise. The helper can draw on past interactions if he or she is counseling a fellow trainee whom he or she has counseled before. At the end of eight minutes or so, the helper should engage in a second summary.
5. At the end of each summary, the helper should ask the client to draw some sort of implication or conclusion from the summary that helps the client move forward.
6. Then both observer and client give the helper feedback as to the accuracy and the helpfulness of the summaries provided. The summaries are useful if they help move the client toward problem clarification, goal setting, and action.
7. This process is repeated until each person in the subgroup has had an opportunity to play each role.

Example 1: It would be too cumbersome to print eight minutes of dialogue here, but consider this brief outline of a case. A young man, 22 years old, has been talking about some developmental issues. One of his concerns is that he sees himself as relating to women poorly. One side of his face is scarred from a fire that occurred two years previous to the counseling session. He has made some previous remarks about the difficulties he has relating to women. After four minutes of interaction, the helper summarizes:

> **HELPER**: "Dave, let me see if I have the main points you've been making. First, because of the scars, you think you turn women off before you even get to talk with them. The second point, and I have to make sure that this is what you are saying, is that your initial approach to women is cautious, or cynical, or maybe even subtly hostile since you've come to expect rejection somewhat automatically."

> **DAVE**: "Yeah, but now that you've put it all together, I am not so sure that the hostility is all that subtle."

> **HELPER**: "So, at least at times, there is some open hostility. . . . You also said that the women you meet are cautious with you. Some might see you as a bit 'nasty,' I believe that was your word. Others steer clear of you because they see you as difficult to talk to and get to know. Finally, what closes the circle is that you interpret their caution as being turned off by your physical appearance."

DAVE: "I don't like to hear it that way, but that's what I've been saying."

HELPER: "If these points are fairly accurate, I wonder what implication you might see in them."

DAVE: "Well, one possibility is that *I'm* the one that rejects me because of my face. If that's the case, then nothing's going to get better until I come to grips with myself."

Note that the client draws an implication from the summary ("I am the primary one who rejects me") and moves on to some minimal declaration of intent ("I need to change this").

Example 2: Artemis, a 37-year-old woman, has been talking about her lack of assertive behavior in a paraprofessional helper training group. She sees clearly that her lack of assertion stands in the way of being an effective helper. She and her helper explore this theme for a few minutes, and then the helper invites her to give a summary of the main points she is making:

HELPER: "Artemis, it might help if you would pull together the main points you've been making. It might help us to see the way forward."

ARTEMIS: "Okay, let's see if I can pull this together. I'm convinced that the ability to move into the life of the client without apologizing for my assertiveness is essential. However, this simply has not been part of my normal interpersonal style. I don't intrude into the lives of others. If anything, I'm too hesitant to place demands on anyone. When I take the role of helper in training sessions, I feel awkward even sharing basic empathic responses and even more awkward using probes. As a result, I let my clients ramble and their problems remain unfocused. In social settings I'm quite passive, except now I'm much more aware of it."

HELPER: "What implication might you draw from it?"

ARTEMIS: "When I put it all together like that, my immediate reaction is to say that I shouldn't try to be a counselor. But I think I would be selling myself short. No matter what career I choose to follow, I need to become more assertive. I have to learn how to take risks with people."

After each trainee gives his or her summary and elicits some reaction from the client, the feedback from the client and the observer should center on the accuracy and the usefulness of the summary, not on a further exploration of the client's problem. Recall that feedback is most effective when it is clear, concise, behavioral, and nonpunitive.

EXERCISE 7-8: HELPING OTHERS ENGAGE IN DIALOGUE

The skills of probing and summarizing can be used to assist hesitant clients become partners in the helping dialogue. Summarizing lets others know you are listening to them in a thoughtful manner. By itself, this encourages clients to open up. Probes both direct the client's attention to important matters *and* indicate the interest of the helper. Combined summarizing and probing can help a client, even a hesitant client, become a fuller partner in the dialogue.

1. Identify people you know who are not good communicators. That is, people who are not effective at getting their story across to others. Limit this to one or two individuals you see on a regular basis.
2. Describe to your learning partner who these individuals are and how you might improve their communication if you were to employ the skills of summarizing and probing.

3. In the next week find an opportunity to engage them in conversation, and make your best effort to listen, summarize, and probe.
4. Report back to your learning partner on the experience. What worked? What didn't work? What impact did you have on the other person?

Chapter 8

STAGE I: TASK 2
FACILITATE CLIENT SELF-CHALLENGE:
FROM NEW PERSPECTIVES TO NEW BEHAVIOR

Chapter 8 in *The Skilled Helper* discusses the role challenging your clients plays in the helping process. Helping clients challenge themselves, both to participate actively in the helping process and to act on what they learn, adds great value. However, while challenge without understanding is abrasive, understanding without challenge is anemic. Before doing the following exercise, read Chapter 8.

We begin with an exercise in self-challenge. Many of the clients you see will be struggling with developmental concerns, similar to those you have struggled with or are currently working on. Like you, they will have both strengths and "soft spots" in coping with the developmental tasks of life.

EXERCISE 8-1: NEW PERSPECTIVES ON BASIC DEVELOPMENTAL TASKS

As mentioned in Chapter 4, many clients struggle with unresolved developmental tasks over their life span. In this exercise we revisit the developmental areas outlined in Chapter 4, but now from the perspective of self-challenge.

1. Review the developmental "soft spots" you identified in Exercise 4-1 (Chapter 4).
2. Pick two of the developmental areas in which you had soft spots. Try to pick areas that will make a difference in your work as a counselor.
3. Challenge whatever blind spots you might have in these areas and come up with one or two new perspectives in each area.
4. Outline the kinds of actions these new perspectives might drive.

Example. Mickey, a 20-year-old undergraduate student majoring in psychology, had this to say:

> **Developmental areas**: "I chose identity and investment in the wider community."
> **New perspective**: "There is a huge difference between identity and what I now see as individualism. I think I've been on the selfish, me-first, what's-in-it-for-me track. I think little about the wider community and belonging to it."
> **Action implications**: "First of all, I have to clear up in my mind the difference between knowing who I am, what I stand for, and where I'm going-- versus my 'me-first' attitude. Second, I'd like to begin to take a critical look at the messages the society of which I am a part is screaming at me through music, movies, television, peer interaction, the internet — the whole package. A lot of the messages that bombard me could be influencing me to remain an adolescent, but I don't want to remain an adolescent."

Now explore two areas for yourself.

First developmental area.

_____

New perspectives.

_____

_____

_____

Link these new perspectives to action.

_____

_____

_____

_____

_____

_____

Second developmental area.

_____

New perspectives.

_____

_____

_____

Link these new perspectives to action.

_____

_____

_____

Share your findings with a partner from your training group.

EXERCISE 8-2: IDENTIFYING DIFFERENT AREAS NEEDING CHALLENGE

Challenge focuses on dysfunctional mindsets, patterns of emotional expression, and behavior, whether internal or external. There are various ways in which clients need help challenging themselves, in order to develop the kinds of new perspectives that lead to constructive problem-managing and opportunity-developing action. In this exercise you are asked to do the following:

87

1. Read the case and determine in what area and in what way this client might benefit from challenge.
2. Sketch out briefly what some helpful new perspectives might look like.
3. Indicate what kind of constructive actions might flow from the new perspective.
4. Share your observations with a learning partner. Note the similarities and the differences among your suggestions. Discuss how you might proceed with each client. Discuss ways of helping these clients challenge themselves.

Example: Lily, a new counselor working in a hospital setting for a number of months, has begun to confide in one of her colleagues-- someone who has worked in health-care facilities for about seven years. Lily has intimated, however tentatively, that she feels that some of the medical doctors tend to dismiss her because she is new, because she is a counselor and not a medical practitioner, and because she is a woman. One doctor has been particularly nasty, even though he masks his abusiveness under his brand of "humor." One day her colleague says to her, "Don't take that stuff. Tell 'em what you think!" Lily replies: "That will only make them worse. I think it more important to stand up for my profession by just being who I am and helping patients as much as possible. Their meanness is a punishment in itself. Anyway, it's a game that doctors play with everyone else in this hospital. That kind of thing has been going on forever. It's not such a big deal anyway."

What mindsets and behavior might be challenged here? Under the guise of "reasonableness," Lily could be into excuse-making. "This is the way it is and the way it will always be" seems to be her stance. Furthermore, Lily says nothing about the possible impact of hospital politics on patients.

What would some helpful new perspectives look like? Lily might be helped to see that accepting the status quo goes counter to the kind of professionalism that should characterize the hospital, and that she prides herself in. She may also be helped to see that the hospital's caste system might interfere with team approaches to helping patients.

What actions might such new perspectives lead to? Lily could bring the core issues before the hospital senate, or she could quietly lobby some of her colleagues to see what kind of consensus there is. She could also talk to the especially abusive doctor privately, and let him know how she feels, or she could confide in one of the doctors who appears free of these prejudices in order to see what course of action she might follow, etc.

Now we turn to a range of clients. It goes without saying that the best way of doing this exercise is through actual interaction with these clients. At this stage, however, you are being asked to do some educated guessing.

1. Dan is in a coaching/counseling session with his supervisor, June. A project is behind schedule and a key customer is complaining. Dan was chosen to be the project leader because he had manifested "leadership" qualities in his work. He says something like this: "Well, this is one of the first team-focused projects we've had. We're not used to working like this. The key people in design and engineering have been slow to come to the table. They keep telling me that they can't let other projects slip. I know the customer wants the prototype by the end of the month, but that's his schedule, not necessarily reality. He's so pushy. Lately, I find it a real chore just talking to him."

What mindsets, behavior, and emotional expression might be challenged here?

_____

_____

_____

What would some helpful new perspectives look like?

_____

_____

_____

What actions might such a perspective lead to?

_____

_____

_____

2. Claudette and Jim, a couple, have tried unsuccessfully to have children. Doctors have examined both of them and now offer little hope even with new medical techniques. They have both been seeing a counselor, but in this session Claudette, who is much less reconciled to their fate than Jim is, is seeing the counselor alone. She says, "It's just not fair. Doctors talk about infertility as a clinical phenomenon, not a human reality. Jim and I love one another. We want to have children. We'd make very good parents. We've done everything the doctors have told us to do, and still nothing."

What mindsets, behavior, and emotional expression might be challenged here?

_____

_____

_____

What would some helpful new perspectives look like?

_____

_____

_____

What actions might such a perspective lead to?

_____

_____

_____

3. Len, married with three teenage children, lost all of his family's savings in a gambling spree. Under a lot of pressure from both his family and his boss, he started attending Gamblers Anonymous meetings. He

seemed to recover-- he stopped betting on horses and ball games. A couple of years went by and Len stopped going to the GA meetings because he "no longer needed them." He even got a better job and was recovering financially quite well. But then his wife began to notice that he was on the phone a great deal with his broker. She confronted him and he reluctantly agreed to a session with one of the friends he made at GA. He says, "She's worried that I'm gambling again. You know, it's really just the opposite. When I was gambling I was financially irresponsible. I bet our future on horses and ball games. But now I'm taking a very active part in creating our financial future. Every financial planner will tell you that investing in the market is central to sound financial planning. I'm a doer. I'm taking a very active role."

What mindsets, behavior, and emotional expression might be challenged here?

_____

_____

_____

What would some helpful new perspectives look like?

_____

_____

_____

What actions might such a perspective lead to?

_____

_____

_____

4. Stefan, a refugee who had been brutalized by a political regime in his native country, has been seeing a counselor in a center that specializes in helping victims of torture. He has been in the country for two years. He has a decent job. Even though he is now secure, at least in some sense of that term, he has found it very difficult to establish relationships with other refugees, with members of the immigrant community, or with native-born Americans. He intimates that he is quite lonely, but dismisses loneliness as "inconsequential" when the counselor brings it up. In one session, he says: "Let's not talk about my so-called loneliness when the world is filled with tragic loneliness. The loneliness that brutality creates, now that's something that counts. I fill my days quite well. When it is time, I'll seek out others, but I need people who can see the world as I see it. As it really is."

What mindsets, behavior, and emotional expression might be challenged here?

_____

_____

_____

What would some helpful new perspectives look like?

_____

_____

_____

What actions might such a perspective lead to?

_____

_____

_____

CHALLENGING NORMATIVE DISHONESTIES

Most of us face a variety of self-defeating dishonesties in our lives. We all allow ourselves, to a greater or lesser extent, to become victims of our own prejudices, smoke screens, distortions, and self-deceptions.

EXERCISE 8-3: CHALLENGING NORMATIVE DISHONESTIES IN ONE'S OWN LIFE

In this exercise, you are asked to confront some of the ordinary ways in which we can be dishonest with ourselves and others. You are asked, as usual, to focus on the kind of dishonesties that might affect the quality of your helping, or the quality of your membership in the training group.

Example 1. In this example a trainee confronts her need to control situations. She says, "I am very controlling in my relationships with others. For instance, in social situations, I manipulate people into doing what I want to do. I do this as subtly as possible. I find out what everyone wants to do and then I use one against the other, and gentle persuasion, to steer people in the direction in which I want to go. In the training sessions I try to get people to talk about problems that are of interest to me. I even use empathic responses and probes to steer people in directions I might find interesting. All this is so much a part of my style that usually I don't even notice it. I see this as selfish, but yet I experience little guilt about it."

Example 2. This trainee confronts his need for approval from others. He says, "Most people see me as a 'nice' person. Part of this I like, but part of it is a smoke screen. Being nice is my best defense against harshness and criticism from others. I'm cooperative. I compliment others easily. I'm not cynical or sarcastic. I've gotten to enjoy this kind of being 'nice.' I find it rewarding. But it also means that I seldom talk about ideas that might offend others. My feedback to others in the group is almost always positive. I let others give feedback on mistakes. Outside the group I steer clear of controversial conversations. But I'm beginning to feel very bland."

Now confront two of your own "dishonesties," which, if dealt with, will help you be a more effective trainee and counselor. In the description section be as specific as you can. As in the examples, describe specific experiences, behaviors, feelings, points of view, and decisions. Give brief examples.

Issue # 1

_____

_____

Describe your "dishonesties."

_____

_____

_____

_____

_____

_____

_____

_____

What should you do about this?

_____

_____

_____

Issue # 2.

_____

_____

Describe your "dishonesties."

_____

_____

_____

_____

_____

_____

_____

What should you do about this?

_____

_____

_____

BLIND SPOTS

Interestingly enough, when we are asked to take a look at our blind spots, they are no longer as blind. But most people don't ordinarily ask themselves, "What are my blind spots?" Read the section on blinds spots in Chapter 8 before doing these exercises.

EXERCISE 8-4: EXPLORING YOUR OWN BLIND SPOTS

We all have blind spots, but some are more damaging than others. In these exercises, try to focus on blind spots that may interfere with your competence as a helper.

1. Name a problem situation and an opportunity-development situation in your life which you did not address (or did not address as well as you could have) because of blind spots.
2. Review the following two examples, and then move on to real situations in your own life.

Problem-Focused Example: Lunetta came from a very strict family. Growing up was not at all unpleasant for her, because she easily took to developing the kind of self-discipline that was expected of her. She and her two brothers studied hard, did their share of household work, attended church, and volunteered in the community. Brothers and sister got along well and cooperated to get things done. She grew up believing that if everyone didn't live this way, they *should*. At school and in the community she tended to become friends with like-minded young people. For the most part they avoided the "nonsense" that went on in school and in the community. She was pretty much unaware of what the world was really like. Her definition of "normal" was very restricted. At the end of high school, she decided to get a job for a year or two before moving on to college. She got a job in a large retailer. The store she worked in was the anchor store in a suburban mall. At the store she met all sorts of young people, both coworkers and customers. Because so many of them did not seem to live up to her standards, she felt disorganized and frustrated. She got into a number of arguments with other workers, and even with younger customers. On the verge of being fired, she quit. At home, she became listless and sullen.

What blind spots might she have had, and what possible part do you think they played in the difficulties she was experiencing?

_____

_____

_____

_____

Opportunity-Focused Example: Craig, a single man, retired on a modest company pension when he was 62 years old. He also opted for the reduced social security pension that was still available for 62-year olds. These two sources of income enabled him to live modestly but comfortably. He owned the condominium in which he lived. Craig was an "avoider." Whenever there were problems at work, when people got in his hair, he daydreamed about how "great" retirement would be. He did little to prepare psychologically for retirement. Within a month he missed the social life at work and also the intellectual challenges of his job. He realized that he was still comparatively young. He lived in a city with a lot of amenities. His health was good. But three months after retiring, surrounded by opportunities, he was miserable.

What part did blind spots possibly play in his predicament?

_____

_____

_____

_____

Describe a problem situation in your own life you managed poorly because of blind spots.

_____

_____

_____

Describe an unused opportunity in your own life you managed poorly because of blind spots.

_____

_____

_____

EXERCISE 8-5: HELPING OTHERS IDENTIFY AND DEAL WITH BLIND SPOTS

In this exercise you are asked to search your past and present relationships and interactions with others—relatives, friends, teachers, acquaintances, coworkers, and so forth—in order learn from blind spots you have observed in them. This exercise deals with different levels of unawareness and tendencies to distort the world instead of seeing it as it is. Review the "Degrees of Awareness" section in Chapter 8 before doing this exercise. You are looking for blind spots that have caused or are now causing problems for the person. Blind spots deal with ways of thinking, modes of emotional expression, behavior, or a combination of all of these.

1. **"Simple unawareness" example**. Beth seems totally unaware that her failure to control the behavior of her two small children annoys most of her friends, and is not in the best interest of the children.

Describe a case of simple unawareness from your experience.

_____

_____

_____

Describe the consequences of the blind spot.

_____

_____

_____

What could you do or have done to help the person become aware?

_____

_____

_____

2. **"Failure to think things through" example**. Cheryl and Sam get into trouble because they buy a house they like with an interest-only mortgage, without considering the comparative price of real estate in the neighborhood, the state of the real estate market, the state of the economy, interests rates, the likelihood of their wanting to move later on, and so forth.

Describe a case of failing to think things through.

_____

_____

_____

Describe the possible or actual consequences.

_____

_____

_____

What could you do or have done to help the person become aware?

_____

_____

_____

3. **"Self-deception" example.** Rebecca is deceiving herself when she says that she can stop using drugs without changes in her social life, even though she hangs around with a group of people who are serious drug users.

Describe a case of self-deception.

_____

_____

_____

Describe the possible or actual consequences.

_____

_____

_____

What could you do or have done to help the person challenge his or her self-deception?

_____

_____

_____

4. **"Choosing to stay in the dark" example**. Sometimes, Lisa feels that her husband is dissatisfied with their relationship. Both are in their mid-30s and have full-time jobs. She notes when he hints at this possibility,, but then puts it out of her mind. She does not want to deal with it. She also refuses to think about any kind of possible linkage among the various incidents.

Describe a case of choosing to stay in the dark.

_____

_____

_____

Describe the possible or actual consequences.

_____

_____

_____

What could you do or have done to help the person challenge his or her avoidance?

_____

_____

_____

5. **"Knowing, not caring, and ignoring possible consequences" example**. Fernando does not like his boss, and lets her know this in a variety of not-too-subtle ways. He knows that this annoys her, but in the back of his mind thinks that no single incident merits any kind of retaliation. His coworkers watch this little drama. Fernando knows that others are watching and takes pride in getting away with it.

Describe a case of knowing, not caring, ignoring possible consequences.

_____

_____

_____

Describe the possible or actual consequences.

_____

_____

_____

What could you do or have done to help the person challenge his or her behavior?

_____

_____

_____

SPECIFIC SKILLS FOR HELPING CLIENTS ENGAGE IN
SELF CHALLENGE

Specific challenging skills, as outlined in Chapter 8 of *The Skilled Helper*, include sharing advanced empathic responses, information sharing, helper self-disclosure, suggestions, recommendations, confrontation, and encouragement. These skills help clients get in touch with their blind spots, develop the kind of new perspectives or behavioral insights needed to complete the clarification of a problem situation or opportunity, and move on to developing possibilities for a better future, setting problem-managing goals, developing strategies, and moving to action. Challenging skills serve the entire helping process.

SHARING ADVANCED EMPATHIC RESPONSES

Advanced empathy means sharing educated **hunches** about clients and their overt and covert experiences, behaviors, and feelings that you think will help them see their problems and concerns more clearly, as well as help them move on to developing new scenarios, setting goals, and acting. Such hunches, of course, must be based on the helper's interactions with the client, in which active listening, understanding, empathy, and probing play a large part.

Hunches should be based on your experience with and knowledge of your clients—what is happening in their lives, their behaviors and emotions, their points of view, the decisions they have made or are making, and their intended courses of action—both within the helping sessions themselves and in their day-to-day lives. Base you hunches on your relationship with your clients. Be hesitant to base your hunches on "deep" and speculative psychological theories. Be aware of the experiential and behavioral clues on which your hunches are based.

EXERCISE 8-6: ADVANCED ACCURATE EMPATHY—HUNCHES ABOUT YOURSELF

One way to get an experiential feeling for advanced empathy is to explore *at two levels* some situation or issue in your own life that you would like to understand more clearly. One level of understanding could be called the surface or immediate level. The second could be called a deeper or more thoughtful level. Often enough reality is found at the more thoughtful level.

Read the following examples.

Example 1: A 25-year-old man, in a counselor training program, is having some ups and downs in the training group. He is having second thoughts about his ability and willingness to get close to others.

> **Level-1 understanding:** "I like people and I show this by my willingness to work hard with them. For instance, in this group I see myself as a hard worker. I listen to others carefully and I try to respond carefully. I see myself as a very active member of this group. I take the initiative in contacting others. I like working with the people here."

> **Below-the-surface understanding**: "If I look closer at what I'm doing here, I realize that underneath my hardworking and competent exterior, I am uncomfortable. I come to these sessions with more misgivings than I have admitted, even to myself. My hunch is that I have some hesitation about human closeness. I am afraid that both here and in a couple of relationships outside the group someone is going to ask me for more than I want to give. This keeps me on edge here. It keeps me on edge in a couple of relationships outside."

Example 2: A 33-year-old woman, in a counselor training group, sees that her experience in the group is making her explore her attitude toward herself. It might not be as positive as she thought. She sees this as something that could interfere with her effectiveness as a counselor.

> **Level-1 understanding:** "I like myself. I base this on the fact that I seem to relate freely to others. There are a number of things I like specifically about myself. I'm fairly bright. And I think I can use my intelligence to work with others as a helper. I'm demanding of myself, but I don't place unreasonable demands on others."

> **Below-the-surface understanding**: "If I look more closely at myself, I see that when I work hard it is because I feel I have to. My hunch is that I-have-to counts more in my motivation than I-want-to. I get pleasure out of working hard, but it also keeps me from feeling guilty. If I don't work hard enough as I define the term, then I can feel guilty or down on myself. I am beginning to feel that there is too much of the I-must-be-a-perfect-person in me. This leads to my judging both myself and others more harshly than I should."

1. Choose some issue, topic, situation, or relationship that you have been investigating, and which you would like to understand more fully, with a view to taking some kind of action on it. Try to choose issues that might affect the quality of your counseling.
2. Briefly describe the issue, as in the examples.
3. Then give a "surface-level" description of the issue.
4. Next, share some hunch you have about yourself that relates to that issue, a hunch that "goes below the surface," as it were, but don't try to "psych yourself out." This will help you get in touch with possible blind spots. Try to develop a new perspective on yourself and that issue, one that might help you see the issue more clearly, so that you might begin to think of how you might act on it.
5. Share you insights with your learning partner. Discuss the action implications of these insights.

Issue # 1.

_____

_____

Level-1 understanding.

_____

_____

_____

Below-the-surface understanding.

_____

_____

_____

_____

_____

_____

Issue # 2.

_____

_____

Level-1 understanding.

_____

_____

_____

Below-the-surface understanding.

_____

_____

_____

_____

_____

EXERCISE 8-7: ADVANCED VERSUS BASIC EMPATHIC RESPONSES

1. In this exercise, assume that helpers and clients have established a good working relationship, that clients' concerns have been explored from their perspective, and that the clients need to challenge themselves to see the problem situation from some new perspective, or to act on the insights that have been developed.

2. In each instance, imagine each client speaking directly to you.

3. First respond to what the client has just said with a basic empathic response.

4. Next, formulate one or two hunches about this person's experiences, behaviors, feelings, point of view, decision, or intended course of action—hunches that, when shared, would help promote the kind of new perspectives that drive problem-managing or opportunity-developing action. Use the client's statement and context to formulate your hunches. Ask yourself: "On what clues am I basing this hunch?"

5. Finally, respond with some form of advanced empathy by sharing a hunch that you believe will be useful for them.

100

Example: A 48-year-old man, a husband and father, is exploring the poor relationships he has with his wife and children. In general, he feels that he is the victim, that his family is not treating him right (like many clients, he emphasizes his experience, rather than his behavior). He has not yet examined the implications of the ways he behaves toward his family. At this point he is talking about his pride in being a breadwinner. He says: "I work really hard to be a good provider. In fact, I'm more than a provider. I travel a lot and don't really like traveling. But this means that I can provide not just the basics but little luxuries. You know, nice holidays and things like that. But I think I'm more appreciated at work than at home. They see how hard I have to work. Often enough I have to spend hours in my home-office to prepare for my trips. For all this I get a pretty grumpy group at times."

> **Basic empathic response.** "It's irritating when your own family doesn't seem to appreciate what you're doing for them, the sacrifices you're making to give them a better life."
> **Hunch.** The family wants a husband and father, not just a breadwinner or gift-giver. He has unilaterally made a decision to work long hours in order to be the kind of provider he wants to be.
> **Advanced empathic response.** "It might be possible that their 'grumpiness' is not a sign of ingratitude, but frustration because you've chosen not to be around very much."

Now try your hand at the following cases.

1. Clayton, a first-year graduate student in engineering, has been exploring his disappointment with himself and with his performance in school. His father is a successful engineer, but has not pressured his son to follow in his footsteps. Clayton has explored with his counselor such issues as his dislike for the school and for some of the teachers. He says: "I just don't have much enthusiasm. My grades are just okay, maybe even a little below par. I know I could do better if I wanted to. I don't know why my disappointment with the school and some of the faculty members can get to me so much. It's not like me. Ever since I can remember—even in primary school, when I didn't have any idea what an engineer was— I've wanted to be an engineer. Theoretically, I should be as happy as a lark because I'm in a graduate school with a good reputation, but I'm not."

Basic empathic response.

_____

_____

_____

Your hunch and your reason for it.

_____

_____

_____

_____

Advanced empathic response.

_____

_____

_____

_____

2. A man, who is now 64-years-old, retired early from work-- when he was 60 years old. He and his wife wanted to take full advantage of the "golden" years. But, his wife died a year after he retired. At the urging of friends, he has finally come to a counselor. He has been exploring some of the problems his retirement has created for him. His two married sons live with their families in other cities. In the counseling sessions he has been alternately dealing the theme of loss and the theme of redefining his golden years. He says: "I seldom see the kids. I enjoy them and their families a lot when they do come. I get along real well with their wives. But, since my wife has been gone, I don't make the effort I should to make it happen. I have a standing invitation from the boys and just recently I've decided I'm going to get off my sofa and start living again. I won't kid you; it will be bittersweet. I dread those times when I'll want to turn to her and enjoy the moment and she won't be there. I don't want my boys to see their father shattered and I sure as hell don't want to see pity in their eyes."

Basic empathic response.

_____

_____

_____

Your hunch and your reason for it.

_____

_____

_____

_____

Advanced empathic response.

_____

_____

_____

_____

3. A 33-year-old single woman is talking to a psychiatrist about the quality of her social life. She has a very close friend, Ruth, on whom she has become somewhat dependent. She is exploring the ups and downs of this relationship. This is the third session. During the sessions, she comes on a bit loud and somewhat aggressive. She says: "Ruth and I are on again off again with each other lately. When we're on, it's great. We have lunch together, go shopping, all that kind of stuff. But sometimes she seems to click off. You know, she tries to avoid me. But that's not easy to do. *(She laughs.)* I keep after her. She's been

pretty elusive for about two weeks now. I don't know why she runs away like this. Something must be bothering her. I know we have our differences. But we always get over them."

Basic empathic response.

_____

_____

_____

Your hunch and your reason for it.

_____

_____

_____

Advanced empathic response.

_____

_____

_____

4. A 35-year-old divorced woman, who has a 16-year-old daughter, is talking to a counselor about her current relationship with men. She mentions that she has lied to her daughter about her sex life. She told her that she doesn't have sexual relations with men, but she does. In general she seems quite protective of her daughter. She does not know for sure if her daughter is sexually active but she has the feeling the day is not far off when she will start having sex. She says, "I guess I've been afraid that if I told her I was sexually involved that I would lose my authority. How can I tell her to wait until she's married when I'm having sex outside of marriage? And, if I were honest, how much would I have to tell her? Wait. Maybe I can be more honest with her about what I believe without needing to detail my own life. Really this is about how much I love her, not a tell-all TV show. I've wanted to connect with her on this and I think this might be the way. Sometimes, though, it feels like such a risk. What if it goes wrong?"

Basic empathic response.

_____

_____

_____

Your hunch and your reason for it.

_____

_____

_____

_____

Advanced empathic response.

_____

_____

_____

_____

INFORMATION-SHARING

As noted in the text, sometimes clients do not get a clear picture of a problem situation, or manage it effectively, because they lack information needed for clarity and action. Information can provide clients with some of the new perspectives they need to see problem situations as manageable or to spot and embrace unused opportunities. Remember that giving clients problem-clarifying information or, better, helping them find it themselves, is not the same as advice-giving or preaching.

EXERCISE 8-8: INFORMATION LEADING TO NEW PERSPECTIVES AND ACTION

In this exercise you are asked to consider what kind of information you might give or help clients obtain which would enable them see their problem situations or opportunities more clearly. Information can, somewhat artificially, be divided into two kinds: (a) information that helps clients understand their difficulties better and (b) information about what actions they might take.

Example: Tim was a bright, personable young man. During college he was hospitalized after overdosing on drugs during a bout of depression. He spent six months as an in-patient. He was assigned to "milieu therapy," an amorphous mixture of work and recreation designed more to keep patients busy than to help them grapple with their problems and engage in constructive change. He was given drugs for his depression, seen occasionally by a psychiatrist, and assigned to a therapy group that proved to be quite aimless. After leaving the hospital, his confidence shattered, he left college and got involved with a variety of low-paying, part-time jobs. He finally finished college by going to night school, but he avoided full-time jobs for fear of being asked about his past. Buried inside him was the thought, "I have this terrible secret that I have to keep from everyone." A friend talked him into taking a college-sponsored communication-skills course one summer. The psychologist running the program, noting Tim's rather substantial natural talents together with his self-effacing ways, remarked to him one day, "I wonder what kind of ambitions you have." In an instant Tim realized that he had buried all thoughts of ambition. After all, he didn't "deserve" to be ambitious. Tim, instinctively trusting him, divulged the "terrible secret" about his hospitalization for the first time.

Tim and the counselor, Rick, had a number of meetings over the course of a year. In some of the early sessions, Rick provided him with information that proved challenging in a number of ways. For instance, when it came to employment, the counselor helped Tim find out that:

- he was more intelligent and talented than he realized and, therefore, he was underemployed;
- that his hospitalization was quite different from, say, a felony conviction;
- deep background checks were not required for the kinds of jobs that Tim might apply for;
- privacy laws often protect prospective employees from damaging disclosures;
- enlightened employers look benignly on the developmental crises during a prospect's adolescence;
- the job market was strong and growing stronger so that people with Tim's skills and potential were in high demand.

All this challenged Tim to look at himself in a different light-- as a potential winner instead of a loser-- and to move more aggressively into job hunting.

In the following cases, you are not asked to be an information expert. Rather, from a lay person's or a common-sense point of view, you are asked to indicate what kinds of information you believe might help the client understand and manage the problem situation more clearly. What blind spots do you think the client might have? How do you think information might help? In what ways would the information be challenging?

1. Point out the blind spots the client might have.
2. Indicate what information could help the client get some new perspectives and move to action.

1. A 26-year-old man has just been sentenced to five years in a penitentiary, on a felony burglary charge. He is talking to a chaplain-counselor, who has worked for the past ten years at the penitentiary to which the man has been assigned. The man is somewhat cavalier about spending time in the penitentiary. He spends a lot of time looking for ways to overturn his conviction and is preoccupied with the possibility of early parole. His main regret is that he was caught.

What blind spots might the client have?

_____

_____

_____

What information might help this client develop new perspectives?

_____

_____

_____

_____

In what ways might this information be challenging for the client?

_____

_____

_____

_____

2. A 58-year-old woman has endured an almost year long struggle to find out what has been sapping her energy. Her doctors first thought she was depressed and started her on medication, but she saw little improvement and she had to contend with the side effects. Time passed and she lived in uncertainty. Recently, her doctors arrived at a diagnosis—chronic fatigue syndrome. When they told her, the woman's reaction took them a little by surprise. While she was full of questions about what it all meant, still she seemed relieved and eager to get on with whatever would be required. She said it was such a relief to finally know where she stood.

What blind spots might the client have?

_____

_____

_____

What information might help this client develop new perspectives?

_____

_____

_____

In what ways might this information be challenging for the client?

_____

_____

_____

_____

3. Tim, an 18-year-old, has been smoking marijuana for about three years. He is a fairly heavy user. He has recently received several shocks. His father died suddenly and his steady girlfriend left him. Since he was not especially close to his father, he is surprised by how hard he is hit by the loss. After reading a couple of articles on marijuana use, he developed fears that he has been doing irreversible genetic damage

to himself. His is fearful of giving it up because he thinks he needs it to carry him over this period of special stress and because he fears withdrawal symptoms.

What blind spots might the client have?

_____

_____

_____

What information might help this client develop new perspectives?

_____

_____

_____

_____

In what ways might this information be challenging for the client?

_____

_____

_____

HELPER SELF-DISCLOSURE

Although helpers should be *ready* to make disclosures about themselves that would help their clients understand their problem situations more clearly, they should do so only if such disclosures do not upset their clients, or distract their clients from the work they are doing. Read the passage on helper self-disclosure before doing this exercise.

EXERCISE 8-9: EXPERIENCES OF MINE THAT MIGHT BE HELPFUL TO OTHERS

In this exercise you are asked to review some problems in living or unused opportunities that you feel you have managed or are managing successfully. Indicate what you might share about yourself that would help a client with a similar problem situation. That is, what might you share of yourself that would help the client move forward in the problem-managing process?

Example 1: "In the past I have been an expert in feeling sorry for myself whenever I had to face any kind of difficulty. I know very well the rewards of seeing myself as a victim. I used to fantasize about myself as a victim, as a form of daydreaming or recreation. I think many clients get mired down in their problems because they allow themselves to feel sorry for themselves the way I did. I think I can spot this tendency in others. When I see this happening, I think I could share brief examples from my own experience and then ask clients to see if what I was doing squares with what they see themselves doing now."

107

Example 2: "I have been addicted to a number of things in my life and I see a common pattern in different kinds of addiction. For instance, I have been addicted to alcohol, to cigarettes, and to sleeping pills. I have also been addicted to people. By this I mean that at times in my life I have been a very dependent person, and I found the same kind of symptoms in dependency that I did in addiction. I know a lot about the fear of letting go and the pain of withdrawal. I think I could share some of this in ways that would not accuse or frighten clients or distract them from their own concerns."

1. List two areas in which you feel you have something to share that might help clients who have problems in living similar to your own. Just briefly indicate the area.
2. Next, indicate what you might share.
3. Share these with a learning partner. Give and get feedback on the appropriateness and usefulness of the disclosures. Discuss the conditions under which you would share what you have written.

Area # 1.

_____

_____

What in this area might you share, if appropriate?

_____

_____

_____

_____

_____

Area # 2.

_____

_____

What in this area might you share, if appropriate?

_____

_____

_____

_____

_____

Chapter 9

STAGE I: TASK 3
PARTNER WITH CLIENTS IN CREATING
VALUE IN THEIR LIVES

This chapter has two sections. The first section concludes the work on client self-challenge, focusing on the quality of client participation and collaboration in the helping process. Helping clients challenge themselves is, in the end, a way of helping them add value to their lives. The first section also provides counselors with guidelines to help them challenge wisely. The second section deals with the third task of Stage I of the helping process. This task focuses on helping clients choose to work on issues that have substance to them-- issues that will make a difference in their lives. More broadly, this task (as applied to all the stages and tasks of the helping process) focuses on seeking substance and value throughout the change process.

SECTION I: FURTHER THOUGHTS ON CHALLENGE

This section deals with helping clients challenge themselves to find value by participating as fully as possible in the helping process. Helping is, ideally, a collaborative venture. This section also deals with the wisdom-informed guidelines needed to challenge well.

ENCOURAGING CLIENT PARTICIPATION

Clients add value by owning and participating as fully as possible in the helping process. This section focuses on some of the challenges clients face in trying to participate in the process. Even clients whose good will is beyond doubt can have trouble participating for a variety of reasons. First, clients may lack the skills, especially the communication skills, needed to participate fully. If this is the case, you can use your communication skills to help them overcome obstacles to participation. For instance, if Nestor is having problems with expressing himself, use probes to help "walk" him toward clarity. Second, sometimes clients do not participate fully because they do not understand what the helping process is or how it works. If this is the case, find ways to help them understand the essentials of the helping process, as suggested in Chapter 3 of *The Skilled Helper*. Third, clients are often reluctant to participate fully in the helping process because they are reluctant to change. In helping sessions, clients manifest reluctance in many, often covert, ways. They may talk about only safe or low-priority issues, seem unsure of what they want, benignly sabotage the helping process by being overly cooperative, set unrealistic goals and then use their unreality as an excuse for not moving forward, don't work very hard at changing their behavior, and are slow to take responsibility for themselves. They may blame others or the social settings and systems of their lives for their troubles, and play games with helpers. Reluctance admits of degrees; clients come "armored" against change to a greater or lesser degree. Make sure you become a partner with your clients in overcoming whatever form of reluctance they get mired in. Finally, clients fail to participate because they resist what they see as your heavy-handed approach to change. When clients resist, they are no longer partners, but adversaries. Chapter 4 provides hints on to how to work with a client's reluctance or resistance.

EXERCISE 9-1: OWNING PROBLEMS AND UNUSED OPPORTUNITIES

The more fully clients take responsibility for their problem situations and unused opportunities, the more likely are they to try to do something about them. In this exercise, you are asked to look at the ways that you and someone you know own, or fail to fully own, your problem situations and/or unused opportunities.

Example. Sam, a 75-year-old man, is addicted to pain killers. He uses over-the-counter drugs to get rid of even the slightest pain. Because he is getter older, these pains are now more frequent and so his use of drugs is increasing. What does Sam need to know and do in order to own this problem situation?

1. Review the problem situations and unused opportunities in your own life that you have identified in previous exercises.
2. Pick one problem situation that still persists.
3. Jot down the ways in which you are failing to fully own the problem.

_____

_____

_____

3. Indicate what you need to do in order to fully own the problem or unused opportunity and act on it.

_____

_____

_____

4. Repeat this process with respect to someone you know, who does not own some problem situation enough to do something about it. What does this person need to do in order to move forward? How might you help him or her?

EXERCISE 9-2: STATING PROBLEMS AS SOLVABLE

Often clients present problems as unsolvable. This relieves them of the responsibility of doing something about them. Instead they put the burden on you, the helper. You are supposed to use some kind of professional magic that will dispose of the problem.

Example. Lavern, a single mother, is having difficulty controlling the self-destructive behavior of her teenage daughter. She says something like this: "Things were so different when you and I were young. Today society is drowning in permissiveness. Kids get away with things at school that would have gotten us expelled. Tammie (her daughter) is constantly telling me what other kids and other parents are doing. I just get overwhelmed. She sees me as just old-fashioned, out of it. I feel like I'm always swimming against the tide."

1. Give an empathic response to Lavern.

_____

_____

_____

2. .What could you do to help her restate the problem situation in more solvable terms? What would you say to her to invite her to do so?

_____

_____

_____

3. Do the same for one of your own persistent problems or unused opportunities. In what ways do you think of it, at least covertly, in unsolvable terms? Revisit the problem situation or unused opportunity using solvable language.

IMMEDIACY: EXPLORING YOUR RELATIONSHIP WITH A CLIENT

As noted in the text, your ability to deal directly with what is happening between you and your clients during helping sessions is an important skill. **Relationship** immediacy refers to your ability to review the general history and the current status of your relationship with another person, including clients and the members of your training group. **Here-and-now** immediacy refers to your ability to deal with a particular situation that is affecting the ways in which you and another person are relating right now, at this very moment.

Immediacy is a complex skill. It involves: (1) revealing how you are being affected by the other person; (2) exploring your own behavior toward the other person and its impact; (3) sharing hunches about your behavior toward him or her; (4) sharing hunches about his or her behavior toward you; (5) pointing out your own and the other's discrepancies, distortions, smoke screens, and the like, and (6) inviting the other person to explore the relationship with a view to developing a better working relationship—all through dialogue. For instance, if you see that a client is manifesting hostility toward you in subtle, hard-to-get-at ways, you may: (1) let the client know how you are being affected by what is happening in the relationship (you share your experience); (2) explore how you might be contributing to the difficulty; (3) describe the client's behavior and share reasonable hunches about what is happening (challenge); and (4) invite the client to examine in a direct way what is happening in the relationship. Immediacy involves collaborative problem solving and opportunity development with respect to the relationship itself.

EXERCISE 9-3: RESPONDING TO SITUATIONS CALLING FOR IMMEDIACY

In this exercise a number of client-helper situations calling for some kind of immediacy on your part are described. You are asked to consider each situation and respond with some statement of immediacy. Consider the following example.

111

Situation. This client, a 44-year-old, occasionally makes snide remarks either about the helping profession itself or some of the things that you do in your interactions with him. At times he is cooperative and takes responsibility for himself, but at other times he has an edge and asks you to make decisions for him. He is not beyond comments such as, "Just tell me the best way for me to tell my boss that he's an idiot." He makes remarks about you personally at times, "I bet you've got a lot of friends." Or, when invited to challenge himself, he might say "You just won't let me alone, will you?"

Immediacy response. "Tom, let's stop a minute and explore what's happening between you and me in our sessions. . . . (Tom says, "Uh, oh, here we go!") . . . You take mild swipes at the counseling profession such as, 'I hear people are still trying to find out whether counseling works.' Or at me, like, 'Oh, oh, Mr. Counselor is getting a little hot under the collar.' I've ignored these remarks, but in ignoring them I've become a kind of accomplice. Sometimes we seem to be working like a team. And we get things done. Other times you ask me to make decisions for you — you know, the 'just-tell-me' approach. I've let myself get on edge with you and that's not helping us at all. . . . (Tom says, "You want to call my game, huh?") . . . Tom, I guess that I prefer that we not even play games with each other at all. Perhaps we could spend a little time re-setting our relationship."

1. Critique the approach this counselor takes in the example. What would you change?
2. Share your critique with a learning partner.
3. Consider the following situations and write out an immediacy response that helps the client challenge unhelpful perspectives and actions.
4. Share your responses with a learning partner and give feedback to each other. Together come up with more effective immediacy responses.

1. **The situation.** The client is a person of the opposite sex. You have had several sessions with this client. It has become evident that the person is attracted to you and has begun to make thinly disguised overtures for more intimacy. The person finds you both socially and sexually attractive. Some of the overtures have mild sexual overtones.

Immediacy response.

_____

_____

_____

_____

_____

2. **The situation.** In the first session you and the client, a relatively successful, 40-year-old businessman, have discussed the issue of fees. At that time, you mentioned that it is difficult for you to talk about money, but you finally settled on a fee at the modest end of the going rates. He told you that he thought that the fee was "more than fair." However, during the last few sessions he has been dropping hints about how expensive this venture is proving to be. He talks about getting finished as quickly as possible and intimates that is your responsibility. You, who thought that the money issue had been resolved, find it still very much alive.

112

Immediacy response.

_____

_____

_____

_____

_____

3. **The situation.** The client is a young woman, 22 years old, who is obliged to see you as part of being put on probation for a crime she committed. She is cooperative for a session or two and then becomes quite resistant. Her resistance takes the form of both subtle and not-too-subtle questioning of your competence, questioning the value of this kind of helping, coming late for sessions, and generally treating you like an unnecessary burden.

Immediacy response.

_____

_____

_____

_____

_____

EXERCISE 9-4: IMMEDIACY WITH THE OTHER MEMBERS OF YOUR TRAINING GROUP

This exercise applies to members of an experiential training group. It could be called an immediacy "round robin." The exercise assumes that you have interacted for a number of weeks and have gotten to know one another fairly well.

1. If the group is large, divide up into subgroups of about four trainees per group.
2. Read the example below.
3. On separate paper, write out a statement of immediacy for each of the members of your group.
4. Imagine yourself in a face-to-face situation with each member successively. Deal with real issues pertaining to the training sessions, interactional style, and so forth.
5. In a "round robin," share with each of the other members of the group the statement you have written for him or her. Obviously it would be better to speak it than read it.
6. The person listening to the immediacy statement should reply with an empathic response, making sure that he or she has heard the statement correctly.
7. Then, listen to the immediacy statement the other person has for you, and reply with an empathic response.

8. Finally, engage in a dialogue for a few minutes about the implications of the two statements for both your interpersonal style and for this relationship.

9. Continue with the "round robin" until each person has had the opportunity to share an immediacy statement as well as have a dialogue with every other member.

Example: Trainee A to Trainee B. "I notice that you and I have relatively little interaction in the group. You give me little feedback; I give you little feedback. It's almost as if there is some kind of conspiracy of non-interaction between us. I appreciate the way you participate in the group. For instance, the way you invite others to challenge themselves. You do it carefully but without any apology. I think I refrain from giving you feedback, at least negative feedback, because I don't want to alienate you. I do little to make contact with you. I have a hunch that you'd like to talk to me more than you do, but it's just a hunch. I'd like to hear your side of our story."

With a learning partner, identify the elements of immediacy (self-disclosure, challenge, invitation) in this example. Then move on to the exercise.

THE WISDOM OF CHALLENGING

The way in which you challenge clients is extremely important. You need to challenge others in such a way that they will *respond* rather than *react* to your invitations. Before doing the following exercise, read the "From Smart to Wise: Guidelines for Effective Challenging" section in Chapter 9 of *The Skilled Helper*.

EXERCISE 9-5: EFFECTIVE VERSUS INEFFECTIVE CHALLENGE

In this exercise you are asked to practice on yourself. It goes without saying that you should not practice ineffective challenging on others, even colleagues in your training group.

1. Choose in area in which you feel that you need to challenge yourself. It can relate either to a problem situation or some unused resource or opportunity.

2. Write out a self-challenge statement in which you violate the principles of effective challenging outlined in Chapter 9.

3. Then write a self-challenge statement that embodies these principles.

4. Share both forms of self-challenge with a learning partner and discuss. Point out the flaws in the poor self-challenge. Provide feedback to each other on the quality of the appropriate self-challenge.

Example. This trainee is challenging himself on a certain lack of discipline in his life

> **Poor self-challenge**: "You're disorganized and lazy. Your room and your desk are continually messy and your excuse is that you are so busy. But that's a lame excuse. Your wait until the end of the course to do the required papers and therefore their quality is poor. In some ways it is worse in your interpersonal life. You're always late. Others have to wait for you. It's a way of saying you're more important than others. You're just inconsiderate. Since all of this is so ingrained in you, you're not going to change—unless something drastic happens."

> **Effective self-challenge**: "You do reasonably well in school and your social life is quite decent. However, there is the possibility that both could be even more satisfying. At school you tend to put things off, sometimes to the point that rushing affects the quality of your work. For example, end-of-semester papers. Since you pride yourself on giving your best, putting things off like that is not fair to you. You keep telling yourself it's time to be more organized, but you need to come up with a plan for doing so. What would some of the elements of such a plan be?"

114

EXERCISE 9-6: CHALLENGING YOUR OWN UNUSED STRENGTHS

Challenging unused or partially used strengths, rather than weaknesses, is one of the principles outlined in Chapter 9. In this exercise you are asked to confront yourself with respect to your own unused or underused strengths and resources.

1. Briefly identify some problem situations or unused opportunities that you have.
2. As in the example, indicate how some unused or underused strength or strengths you have could be applied to the management of each problem situation or opportunity.
3. In addition, indicate some actions you might take to make use of underused strengths.

Example 1. Katrina, a trainee in her early 20s, is concerned about bouts of anxiety. She has led a rather sheltered life and now realizes that she needs to "break out" in a number of ways if she is to be an effective counselor. The counseling program has put her in touch with all sorts of people, and this is anxiety-provoking. She lists some of the strengths she has that can help her overcome her anxiety.

> **Description of underused strengths or resources and some possible actions**: "I'm bright. I know that a great deal of my anxiety comes from fear of the unknown. I'm intellectually adventuresome. I pursue new ideas. Perhaps this can help me be more adventuresome in seeking new experiences."
>
> **Actions I can take to develop and use these resources**: "The best way to dissipate my fears is to face up to situations that cause them. New ideas don't kill me. I bet new experiences won't either. People find me easy to talk to. This can help me form friendships. My fears have kept me from developing friendships. However, developing more relationships is the best route to the kinds of different experiences I need."

Example 2. This example deals with an unused opportunity. The trainee says, "My social life is not nearly as full as I would like it to be. Extending my social life both quantitatively and qualitatively is an unused opportunity."

> **Description of underused strengths or resources**. "I have problem-solving skills, but I don't apply them to the practical problems of everyday life, such as my less than adequate social life. Instead of defining goals for myself (making acquaintances, developing friendships), and then seeing how many different ways I could go about achieving these goals, I wait around to see if something will happen to make my social life fuller. I remain passive even though I have the skills to become active."
>
> **Actions I can take to develop and use resources**: "First, I am going to get active. I know people who like theater and I'll set up some evenings out. Second, there are a few others I know that are in the same boat I'm in. I think they are as ready to stop complaining about how little there is to do as I am. I'm going to bring the issue up — not to put anyone down but to engage their thinking. And, third, I'm going to make sure that when I enjoy myself on one of these evening's out that I follow-up with other social possibilities."

Now apply the same procedure to one problem situation and one unused opportunity.

1. Problem situation.

_____

_____

115

Underused strengths as applied to the problem situation.

_____

_____

_____

Actions you could take to make use of this underused resource in this problem situation.

_____

_____

_____

2. Unused opportunity.

_____

_____

Underused strengths as applied to the unused opportunity.

_____

_____

_____

Actions you could take to make use of this underused resource to develop this opportunity.

_____

_____

_____

EXERCISE 9-7: TENTATIVENESS IN THE USE OF CHALLENGING SKILLS

As noted in the text, challenges are usually more effective if they do not sound like accusations. Therefore, in challenging clients, don't accuse them, but don't be so tentative that the force of the challenge is lost.

1. Return to Exercise 8-7 on advanced empathy in Chapter 11.
2. With a learning partner, review the responses you wrote for this exercise in terms of tentativeness.
3. Improve responses that would benefit from more, less, or better expressed tentativeness.
4. Share your redone responses with a learning partner and provide mutual feedback.

SECTION II.
STAGE I: TASK 3
HELP CLIENTS WORK ON ISSUES THAT WILL
ADD VALUE TO THEIR LIVES

The exercises in Task 3 deal with helping clients work on issues that will make a difference in their lives. Counseling time is too precious to waste on issues that do not add real value to a client's life. Read Section II of Chapter 9 of *The Skilled Helper* before doing the exercises in this section.

EXERCISE 9-8: HELPING CLIENTS SCREEN PROBLEMS

Clients may need help in determining whether their issues are important enough to bring to a helper in the first place. This is called "screening."

1. Read these two case summaries with a view to discussing them with your fellow trainees.
2. Discuss these two cases in terms of the material in the text on screening. Consider these questions.

> How would you approach the woman in Case 1?
> Under what conditions would you be willing to work with her?
> How would you approach the man in Case 2?
> In what general ways does this case differ from Case 1?
> What would your concerns be in working with the man in Case 2?

Case 1: Christine got pregnant in her first year of college. She had the baby and six months later she and the child's father married. Though she left school briefly, she registered at a local college and managed to re-establish her dream of a college education. She is more mature now than she would ever have been had she followed the path first laid down for her. She's certain of that. The marriage is strong, her child a blessing, and her bachelor's degree is within sight. The question now is what to do next. The career counselor at the school thinks she should consider law school. Friends tell her that she would make an excellent counselor. Christine thinks of having more children, too. Fortunately, there is enough money to realistically consider these options and others. She knows that she has to make some important decisions, so she comes to you for help.

Case 2: Ray, a 41-year-old man, is a middle manager in a manufacturing company located in a large city. He goes to see a counselor with a somewhat complex story. He is bored with his job; his marriage is lifeless; he has poor rapport with his two teenage children, one of whom is having trouble with drugs; he is drinking heavily; he doesn't think much of himself; he has begun to steal things, small things, not because he needs them but because he gets a kick out of it. He tells his story in a rather disjointed way, skipping around from one problem area to another. He is a talented, personable, engaging man who seems to be adrift in life. He does not show any symptoms of severe psychiatric illness. He does experience a great deal of uneasiness in talking about himself. This is his first visit to a helper.

EXERCISE 9-9: CHOOSING ISSUES THAT MAKE A DIFFERENCE

Often enough, the stories clients tell are quite complex. And so they may need your help in deciding which issues to work on first and which merit substantial attention. In this exercise you are asked to review the *personal concerns and problems* you have identified in doing the exercises so far in this manual.

117

1. Briefly list about ten concerns—major, middling, or minor—or unexplored opportunities—large, medium, or small—that you have discovered in doing the self-assessment exercises up to this point.

_____

_____

_____

_____

_____

_____

_____

_____

_____

2. Next, do some screening. Put a line through those that you would probably not bring to a counselor, either because they are not that important, or because you believe that you could handle them easily if you wanted.

3. Review the remaining items on your list in the light of the following "leverage" criteria:

- If there is a crisis, first help the client manage the crisis.
- Begin with the problem that seems to be causing pain for the client.
- Begin with issues the client sees as important and is motivated to work on.
- Begin with some manageable sub-problem of a larger problem situation.
- Begin with a problem that, if handled, will lead to some kind of general improvement in the client's condition.
- Focus on a problem for which the benefits will outweigh the costs.

4. Rank the items on your list. Then share the rankings with a learning partner. Explain the "leverage" reasons behind your rankings. Explain your choices and rankings in terms of the criteria outlined above or any further criteria not listed there. What is the value, the "bang for the buck," of each?

Example: Gino is a trainee in a clinical psychology program. Here is one of the concerns on his list: "I am very inconsistent in the way I deal with people. For instance, some of my friends see me as fickle. I blow hot and cold. One friend told me that whenever he sees me, he's not sure which Gino he will meet. In fact, I seem to be inconsistent in other areas of life." Gino thinks this issue is important, one he has to work on. It is a question of social maturity. He also believes that if he were to handle this issue well, it would spill over into other areas of his life, because some of the same kinds of inconsistency plague him, for instance, in his pursuit of his studies. He runs hot and cold, studying only when he wants. There is a theme here. The benefits of working on the theme will certainly outweigh the costs.

PULLING STAGE I TOGETHER

The following exercise asks you to pull together what you have learned about Stage I of the helping process and apply it to yourself.

EXERCISE 9-10: COUNSELING YOURSELF: AN EXERCISE IN STAGE I

At this point, you have developed an overview of the helping model and, through reading and doing the exercises in this manual you have also developed an understanding and behavioral "feel" for Stage I. In this exercise you are asked to carry on a dialogue with yourself in writing. Choose a problematic area of your life or some unused opportunity—something that is relevant to your success as a helper. First use empathic highlights, probes, and challenges to help yourself tell your story. Work at clarifying it in terms of specific experiences, behaviors, and feelings. Explore your own points of view. Clarify your decisions. Take a closer look at intended courses of action related to the issue you choose. Both probe and challenge yourself with respect to the "little" actions you can take to manage this problem or opportunity.

Example: This example comes from the experience of Cora, a woman in a master's degree in counseling psychology. Here is some of her dialogue with herself.

Her initial story. "To be frank, I have a number of misgivings about becoming a counselor. A number of things are turning me off. For instance, one of my instructors this past semester was an arrogant guy. I kept saying to myself, 'Is this what these psychology programs produce? Could this guy really help anyone?' I also find the program much too theoretical. In a Theories of Counseling and Psychotherapy course we never did anything, not even discuss the practical implications of these theories. And so they remained just that—theories. I'm very disappointed. I'm about to go into my second year, but I have serious reservations. From what others tell me, the program gets a bit more practical, but not enough. There's a practicum experience at the end of the program, but I want more hands-on work now. So I've started working at a halfway house for people discharged from mental hospitals. But that's not working out the way I expected either. There's something about this whole helping business that is making me think twice about myself and about the profession."

SELF-COUNSELOR: All of this adds up to the fact that the helping profession, at least from your experience, is not what it's cracked up to be. The program and some of the instructors in it have left you quite disappointed. Of all these issues, which one or ones hits you the hardest?
SELF-CLIENT: Hmm. It's hard to say, but I think the halfway house bothers me most. Because that's not theoretical stuff. That's real stuff out there.
SELF-COUNSELOR: That's a place where real helping should be taking place. But you've got misgivings about what's going on there.
SELF-CLIENT: Yes, two sets of misgivings. One set about me and one about the place.
SELF-COUNSELOR: Tell me more about one set or the other.
SELF-CLIENT: I feel I have to explore both, but I'll start with myself. I felt so ill-prepared for what I encountered. What's in the lectures and books seems so distant from the realities of the halfway house. For instance, the other day one of the residents there began yelling at me when we were passing in the hallway. She hit me a few times and then ran off screaming that I was trying to get her in some way. I didn't know what to do. I felt so incompetent. And I felt guilty.
SELF-COUNSELOR: That sounds pretty upsetting. You just weren't prepared for it. I'm wondering whether the more practical part of the counseling program you're going into would better prepare you for that kind of reality.
SELF-CLIENT: It could be. I may be doing myself in by jumping ahead of myself.
SELF-COUNSELOR: But you still have reservations about the effectiveness of the halfway house.

SELF-CLIENT: I wasn't ready for what I found there. I've been there a couple of months. No one has really helped me learn the ropes. I don't have an official supervisor. I just see all sorts of people with problems and help when I can.

SELF-COUNSELOR: You just don't feel prepared and they don't do much to help you. So you feel inadequate. You also seem to be basing your judgments about the adequacy of helper-training programs and helping facilities in general, on your experience in this training program, and on the halfway house.

SELF-CLIENT: That's a good point. I'm making the assumption that both should be high-quality places. As far as I can tell, they're not. I guess I have to work on myself first. But that's why I went to the halfway house in the first place. I'm an independent person, but I'm too much on my own there. In a sense, I'm trusted, but, since I don't get much supervision, I have to go on my own instincts and I'm not sure they're always right.

SELF-COUNSELOR: There's some comfort in being trusted, but without supervision you still have a what-am-I-doing-here feeling.

SELF-CLIENT: Right. There are many times when I ask myself just that, "What are you doing here?" I provide day-to-day services for a lot of people. I listen to them. I take them places, like to the doctor. I get them to participate in conversations or games and things like that. But it seems that I'm always just meeting the needs of the moment. I'm not sure what the long-range goals of the place are, and if anyone, including me, is contributing to them in any way.

SELF-COUNSELOR: You seem to have started with a rather difficult population. You get some satisfaction in providing the services you do, but you don't feel equipped to add real value in the halfway house. You may even wonder what this value might look like. Anyway, the lack of overall purpose or direction for yourself and the institution is frustrating. I'm not sure whether you've asked yourself what you might do about all of this, either at the halfway house or in the counseling program.

SELF-CLIENT: I'm letting myself get frustrated, irritated, and depressed. I'm down on myself and down on the people who run the house. It's a day-to-day operation that sometimes seems to be a fly-by-night venture. See! There I go. You're right. I'm not doing anything to handle my frustrations. I'm just blaming everyone else. And I'm doing little to try to better myself. I'm just beginning to put it into some kind of context.

SELF-COUNSELOR: So this is really the first time you've stopped to take a critical look at yourself as a potential helper and the settings in which helping takes place. And what you see is depressing.

SELF-CLIENT: Absolutely.

SELF-COUNSELOR: I wonder how fair you're being to yourself, to the profession, and even to the halfway house.

SELF-CLIENT (Pause): Well, that's a point. I'm speaking — and making judgments — out of a great deal of frustration. I don't want to go off half-cocked. But I could share my concerns about the psychology program with one of the instructors. She teaches the second-year trainees. She gave a very practical talk. I could also see whether my concerns are shared by my classmates. I could also talk to the second-year students to get a feeling for how practical next year might be.

SELF-COUNSELOR: All in all, you can give yourself a chance to get a more balanced perspective.

SELF-CLIENT: Right. I need to do something instead of just brooding and complaining. But there's still the halfway house. I don't want to come across as the wet-behind-the-ears critic.

SELF-COUNSELOR: I'm not sure whether you're assuming that halfway house you're working in should be a state-of-the-art facility?

SELF-CLIENT: Touché! . . . . Wow, I am basing a lot of my feelings about the profession on that place.

SELF-COUNSELOR: Who at the house might you trust to help you get a better perspective?

SELF-CLIENT: Well, hmm... . . . . There is a consultant who shows up once in a while. He runs staff meetings. And he does it in a very practical way. Maybe I can get hold of him and get a better picture.

SELF-COUNSELOR: So, overall?

SELF-CLIENT: I've got some work to do before I rush to judgment. I like the fact that I want the people and the institutions in the profession to be competent. Deep down I think that I'd make a good helper. I say to myself, "You're all right; you're trying to do what is right." Also, I need to challenge my idealism. I spend too much time grieving over what is happening — or not happening — at school and at the halfway house. I need a more mature outlook and I need to figure out how to turn minuses into pluses, including my own.

1. Review this trainee's responses to herself with a learning partner. What kinds of responses did she use? How would you evaluate their quality? What progress did she make? Describe the movement she made during the session. What responses would you have changed?

2. Next choose a problematic area that is important to you—ideally, one involving both problems and unused opportunities—and then, on separate sheets of paper, write out a helping dialogue with yourself like the one above.

3. Exchange dialogues with your learning partner. Critique what you have written, in terms of the process and skills of Stage I outlined and illustrated in Chapters 4 through 8.

PART THREE

THE UNDERAPPRECIATED DIMENSIONS OF HELPING
HELP CLIENTS CRAFT & CREATE A BETTER FUTURE

Stages II and II of the helping process are about helping client review possibilities for a better future, setting goals, and getting things done. Since this is what business people do, a great deal of the innovation in change-management is found there. The helping industry is behind in what it has to offer for Stages II and III. Therefore, I have tried to incorporate some of my findings from the world of organizations, for-profit, non-profit, private, public, small, and large. For instance, the literature about education and educational change is very rich, and there are lots of good ideas. Some of these ideas are implemented, but many aren't. Many of the change programs sponsored by businesses begin well and then fall by the wayside. The darker side of human nature asserts itself, not just with individuals and their problem situations, but in groups and organizations as well. Understanding the darker side of human nature without losing one's balance, sense of humor and deep-rooted optimism is, in my mind, an important part of wisdom.

Chapter 10 introduces Stages II and II by reviewing the place of decision-making, solutions, and goal-setting in the helping process. A great deal of wisdom has been gathered concerning such essentials, but my fear is that only a fraction of this wisdom drives problem-managing behavior. The good news is that this fact creates opportunities and opens up savannahs of hope and possibility. Chapter 11 focuses on helping clients, formally or informally, review possibilities for a better future, set goals, and make a commitment to these goals. Chapter 12 deals with designing programs for accomplishing problem-managing and opportunity-developing goals, together with reviewing the challenges associated with implementing those goals.

In Chapter 13, the book ends with an overview of the Action Arrow. Every Stage and Task of the helping process can serve as a stimulus for problem-managing action. Chapter 13 focuses on the problems and perils of implementing programs and plans.

Chapter 10

INTRODUCTION TO STAGES II AND III: DECISIONS, GOALS, AND PLANS

In many ways, Stages II and III, together with the Action Arrow, are the most important parts of the helping model because they are about "solutions." It is here that counselors help clients develop and implement programs for constructive change. The payoff for identifying and clarifying both problem situations and unused opportunities lies in doing something about them. The skills helpers need to help clients to do precisely that – engage in constructive change – are reviewed and illustrated in Stages II and III. Chapter 10, as an introduction to Stages II and III, focuses on the somewhat strange nature of decision making, the importance of a solution-focused mindset in helping, the value of goal setting, and the role of hope in the therapeutic process. Read Chapter 10 before doing these exercises.

EXERCISE 10-1: YOUR DECISION-MAKING STYLE

Since, as a counselor, you will be helping clients make decisions about a wide range of issues, you can benefit by taking a look at your own decision-making competence and style.

1. After reading the section on rational decision making, list your strengths as a "rational" decision maker. Give some examples.

_____

_____

_____

_____

_____

_____

2. Next, read the section on the shadow side of decision making. In what ways and in what circumstances does your decision making end up "in the shadows"? Give some examples.

_____

_____

_____

_____

_____

_____

123

_____

_____

_____

3. After reading the section on making smarter decisions, indicate what you need to do to become a more effective decision maker. Give one or two examples of decisions you are likely to make in the near future, and what you might do to make these decisions "smart."

_____

_____

_____

_____

_____

_____

EXERCISE 10-2: HOW SOLUTION-FOCUSED ARE YOU?

Consider the way you go about managing your own problems and helping others with theirs. On a scale of 1-7, rate each of the elements that go into a solution-focused approach to problem management. Read what is said about each element. Then, rate yourself and make a short comment on each rating.

1. **Positive philosophy**. Rating: ( )

Comment: _____

2. **View of clients**. Rating ( )

Comment: _____

3. **Focus on past successes instead of failures**. Rating: ( )

Comment: _____

4. **Role of helper when helping others**. Rating: ( )

Comment: _____

5. **Exploring and exploiting competencies, successes, and "normal times."** Rating: ( )

Comment: _____

6. **Manner of talking about problems**. Rating: ( )

Comment: _____

124

7. **Highlighting insight into resources rather than problems**. Rating: ( )

Comment: _____

8. **Constructive dreaming**. Rating: ( )

Comment: _____

9. **Designing solutions**. Rating: ( )

Comment: _____

10. **Getting things done**. Rating: ( )

Comment: _____

Discuss with a partner what you have learned about yourself.

EXERCISE 10-3: THE VALUE OF GOAL SETTING

Too often we take goal setting for granted. We set goals everyday, usually without realizing that we are doing do.

1. Picture yourself talking with a client, Bonita, who is very good at telling her story but avoids going beyond Stage I.
2. Your task is to explain to her how important it is to move from a *discussion* about her problems and opportunities to *managing* them.
3. On a separate piece of paper, write out what you might say to her about the value of Stages II and III. Remember that, in real life, of course, you would do this through dialogue with Bonita.
4. Through dialogue with a learning partner, share what you have written. Give feedback to each other on what you both have written. If you were Bonita, how might you react to what your learning partner is saying in their statement?

EXERCISE 10-4: HOW GOAL-ORIENTED ARE YOU?

If you are to help clients set goals and come up with strategies for accomplishing them, it is helpful to understand your own approach to goal setting.

1. When it comes to goal setting, whether formal or informal, what are your strengths?

_____

_____

_____

_____

2. When it comes to goal setting, whether formal or informal, what are your weaknesses?

_____

_____

_____

_____

_____

3. What do you need to do to become a better goal-setter in your own personal life?

_____

_____

_____

_____

_____

EXERCISE 10-5: SECOND-ORDER VERSUS FIRST-ORDER CHANGE

As noted in Chapter 11 of *The Skilled Helper*, Singhal, Rao, and Pant (2006) highlight the differences between first-order and second-order change as follows:

- Adjustments to the current situation versus changing the underlying system
- Motoring on as well as possible versus creating something new
- Change prone to collapse versus change that is designed to endure
- Shoring up or fixing versus transforming
- Changed based on old learning or no learning versus changed based on new learning
- Current set of values and behaviors staying in place versus a fundamental shift in values and behaviors
- The persistence of an old narrative versus the creation of a new narrative
- Fiddling with symptoms versus attacking causes

Remember that first-order change is not bad and second-order good. Both have their uses. Adjustments to a situation might be more appropriate than trying to change the underlying system. There is a continuum between the two. In this exercise you are asked to review a couple of cases that first appeared in Chapter 8,

126

and give an estimate of the kinds of either first-order or second-order change that might be called for. Of course, your estimates could change if you were talking directly to the client.

Case 1: Joel and Sandra have been married for less than a year, but rumor has it that they are now seeing a counselor. Joel has a permanent limp as a result of a car accident that occurred during his adolescence.. But, he doesn't define himself by this physical limitation. He is both bookish and a free spirit. He will do almost anything on a dare. He is also generous to a fault. He will do anything for a friend. To the casual observer, Sandra has everything. She is bright, beautiful, and engaging. But those who know her well see her as selfish, willful, and in private, outright mean. One of their acquaintances overheard her making fun of her husband's limp. Those who knew both of them prior to their marriage had misgivings about them getting married from the beginning. Those who knew Sandra wondered how she could make a deeper relationship function. Those who knew Joel well feared that he would capitulate to Sandra's "dark side." His persistent optimism, they said, would be his downfall. However, Joel has begun to challenge Sandra's bad behavior. He is uncomfortable in this role, but that doesn't stop him. Until he came along, people have been afraid of challenging Sandra. She has been able to easily dismiss the both the challenger and the challenge. However, she realizes that the current situation with Joel is a "new game" (she says this to herself) and she doesn't know how to play it. One night, while out with friends, a small disagreement kicks off a huge fight. They both end up saying things they regret. One week later, they are seeing a counselor.

What kinds of first-order change might be useful? Why?

_____

_____

_____

_____

What kinds of second-order change might be useful? Why?

_____

_____

_____

_____

Case 2: Here is a case we have seen before. Craig, a single man, retired on a modest company pension when he was 62 years old. He also opted for the reduced social security pension which was still available for 62-year olds. These two sources of income enabled him to live modestly but comfortably. He owned the condominium in which he lived. Craig was an "avoider." Whenever there were problems at work, when people got in his hair, he daydreamed about how "great" retirement would be. He did little to prepare psychologically for retirement. Within a month, he missed the social life at work and also the intellectual challenges of his job. He realized that he was still comparatively young. He lived in a city with a lot of amenities. His health was good. But three months after retiring, surrounded by opportunities, he was miserable.

What kinds of first-order change might be useful? Why?

_____

_____

_____

_____

What kinds of second-order change might be useful? Why?

_____

_____

_____

_____

Case 3: Here is another case we have seen before. Martina, a 27-year-old, asks a counselor in private practice for an appointment to discuss "a number of issues." Martina is both verbal and willing to talk, and her story tumbles out in rich detail. Although the helper uses the skills of attending, listening, sharing highlights, and probing, she does so sparingly. Martina is eager to tell her story.

Although trained as a nurse, Martina is currently working in her uncle's business because of an offer she "could not turn down." She is doing very well financially, but she feels guilty because service to others has always been a value for her. And although she likes her current job, she also feels hemmed in by it. A year of study in Europe during college whetted her appetite for "adventure." She feels that she is nowhere near fulfilling the great expectations she has for herself.

She also talks about her problems with her family. Her father is dead. She has never gotten along well with her mother, and now that she has moved out of the house, she feels that she has abandoned her younger brother, who is 12 years younger than she is, and whom she loves very much. She is afraid that her mother will "smother" her brother with too much maternal care.

Martina is also concerned about her relationship with a man who is 2 years younger than she. They have been involved with each other for about 3 years. He wants to get married, but she feels that she is not ready. She still has "too many things to do" and would like to keep the arrangement they have. This whole complex story—or at least a synopsis of it—comes tumbling out in a torrent of words. Martina feels free to skip from one topic to another. The way Martina tells her story is part of her enthusiastic style. At one point she stops, smiles, and says, "My, that's quite a bit, isn't it!"

What kinds of first-order change might be useful? Why?

_____

_____

_____

_____

What kinds of second-order change might be useful? Why?

_____

_____

_____

_____

Singhal, A., Rao, N., & Pant, S. (2006). Entertainment-education and possibilities for second-order social change. *Journal of Creative Communications*, *1*, 267–283.

Chapter 11

STAGE II: THE PREFERRED PICTURE –
HELP CLIENTS DESIGN A BETTER FUTURE AND COMMIT THEMSELVES TO IT

Problems can make clients feel hemmed in and closed off. They can sometimes feel that they have no future, or the future they have looks troubled. But, as Gelatt (1989) noted, "The future does not exist and cannot be predicted. It must be imagined and invented." (p. 255). The interrelated tasks of Stage II outline three ways in which helpers can partner with their clients with the intention of exploring and developing this better future.

> **Task 1—Possibilities**. "What possibilities do I have for a better future?" "What would my life look like if it looked better?" In helping clients move from problems to solutions, counselors help them develop a sense of hope.
> **Task 2—Choices**. "What do I really want and need? What solutions are best for me?" Here counselors help clients craft a viable change agenda from among the possibilities. Helping them shape this agenda is the central task of helping.
> **Task 3—Commitment**. "What am I willing to pay for what I want?" Help clients discover incentives for commitment to their change agenda. It is a further look at the economics of personal change discussed in Task 3 of Stage I.

TASK 1:
HELP CLIENTS DISCOVER POSSIBILITIES FOR A BETTER FUTURE
"What Would a Better Life Look Like?"

Effective helping requires imagination. In this task you help clients develop a vision of a better future. Once clients understand the nature of the problem situation or identify a key unused opportunity, they need to ask themselves, "What do I want? What would my situation look like if it were better, at least a little bit better?" Read Chapter 11 before doing these exercises.

EXERCISE 11-1: PROBLEMS AND OPPORTUNITIES IN THE SOCIAL SETTINGS OF LIFE

This is another exercise that will help you review areas in which clients have problems and also help you identify your own strengths and weaknesses. Individuals belong to and participate in a number of different social settings in life: family, circle of friends, clubs, church groups, school groups, work groups, and the like. People are also affected by what goes on in their neighborhoods and the cities and towns in which they live. Larger systems, such as state and national governments, also have their ways of entering people's lives.

1. Chart the social settings of your life as in the example below. The person in the example is Mitch, a 45-year-old principal of an inner-city high school in a large city. He is married and has two teenage sons. Neither attends the high school of which he is principal. He is seeing a counselor because of exhaustion and bouts of hostility and depression. He has had a complete physical check-up and there is no evidence of any medical problem.
2. Choose two or three social settings that are currently important in your life. These are settings in which you are experiencing problems or have unused opportunities.

130

3. Review issues, demands, conflicts, concerns. Now take each key social setting and write down issues or concerns in that setting. For instance, some of things Mitch writes are:

School

- Some faculty members want a personal relationship with me and I have neither the time nor the desire.
- Some faculty members have retired on the job. I don't know what to do with them.
- Some of the white faculty members are suspicious of me and distant just because I'm black.
- One faculty member wrote the district superintendent and said that I was undermining her reputation with other faculty members. This is not true.
- The students, both individually and through their organizations, keep asking me to be more liberal, while their parents are asking me to tighten things up.

Family

- My wife says that I'm letting school consume me; she complains constantly because I don't spend enough time at home. Even though she wants me to spend more time at home, she criticizes me for not spending more time with my parents.
- My kids seem to withdraw from me because I'm a double authority figure, a father and a principal.

Parents

- My mother is infirm; my retired father calls me and tells me what a hard time he's having getting used to retirement.
- My mother tells me not to be spending time with her when I have so much to do and then she complains to my wife and my father when I don't show up.

School District

- I would like to become the district superintendent. I think I could do a lot of good.

Friends

- My friends say that I spend so much time at work involving myself in crisis management that I have no time left for them; they tell me I'm doing myself in.

2. Review possibilities for a better future. Choose two concerns arising from two different social settings of your life. Spell out some of the possibilities for a better future in this area. What would this problem or unused opportunity look like if it were managed better or solved? What are some of the things that would be in place that are not now in place? Use the following probes to help yourself brainstorm possibilities.

- Here's what I need . . . .
- Here's what I want . . . .
- Here are some items on my wish list . . . .
- When I'm finished I will have . . . .
- There will be . . . .
- I will have in place . . . .
- I will consistently be . . . .
- There will be more of . . . .
- There will be less of . . . .

For instance, Mitch, in reviewing the conflict between his work and his friends, comes up with these possibilities: more time with his friends, greater flexibility in managing his calendar at work, better integration of his friendships with his home life, a more clear-cut division between his work setting and the other social settings of his life, more time for himself, a clearer understanding of his career aspirations and the costs associated with them, and better delegation on his part to the members of the school administrative team. Now do the same for yourself in two or three of the social settings of your life on a separate sheet of paper.

First social setting.

_____

Social-setting related issue, demand, conflict, opportunity, concern.

_____

_____

Possibilities for a better future in this area of concern.

_____

_____

_____

_____

Second social setting.

_____

Social-setting related issue, demand, conflict, opportunity, concern.

_____

_____

Possibilities for a better future in this area of concern. _____

_____

_____

_____

_____

_____

Third social setting.

_____

Social-setting related issue, demand, conflict, opportunity, concern.

_____

_____

Possibilities for a better future in this area of concern.

_____

_____

_____

_____

_____

EXERCISE 11-2: BRAINSTORMING POSSIBILITIES FOR A BETTER FUTURE—CASES

This exercises focuses on a number of cases, many of which you have seen before. Now you are asked to put yourself in each person's shoes and brainstorm possibilities for a better future. What would things look like if they looked better? If you were this person, what are some of the things you would want to replace what you have?

1. Read the case. Put yourself in this person's shoes. Make his or her problem situation your own.
2. Point out a blind spot this person might have-- a blind spot that might keep this person from managing the problem situation or developing an opportunity more effectively.
3. Brainstorm a number of possibilities for a better future. What would you want if you were this person?
4. Use the probes outlined in the previous exercise to develop possibilities for a better future.

1. Tim was a bright, personable young man. During college, he was hospitalized after overdosing on drugs during a bout of depression. He spent six months as an in-patient. He was assigned to "milieu therapy," an amorphous mixture of work and recreation, designed more to keep patients busy than to help them grapple with their problems and engage in constructive change. He was given drugs for his depression, seen occasionally by a psychiatrist, and assigned to a therapy group that proved to be quite aimless. After leaving the hospital, his confidence shattered, he left college and got involved with a variety of low-paying, part-time jobs. He finally finished college by going to night school, but he avoided full-time jobs for fear of being asked about his past. Buried inside him was the thought, "I have this terrible secret that I have to keep from everyone." A friend talked him into taking a college-sponsored communication-skills course one summer. The psychologist running the program, noting Tim's rather substantial natural talents together with his self-effacing ways, remarked to him one day, "I wonder what kind of ambitions you have." In an instant Tim realized that he had buried all thoughts of ambition. After all, he didn't "deserve" to be ambitious. Tim, instinctively trusting the program director, divulged the "terrible secret" about his hospitalization for the first time.

A key blind spot translated into a future-oriented new perspective.

_____

_____

A range of possibilities for a better future.

_____

_____

_____

_____

2. Cormack is in a master's degree program in counseling psychology. He says: "To be frank, I have a number of misgivings about becoming a counselor. A number of things are turning me off. For instance, one of my instructors this past semester was an arrogant guy. I kept saying to myself, 'Is this what these psychology programs produce? Could this guy really help anyone?' I also find the program much too theoretical. In a 'Theories of Counseling and Psychotherapy' course we never did anything, not even discuss the practical implications of these theories. And so they remained just that—theories. I'm very disappointed. I'm about to go into my second year, but I've got serious reservations. From what others tell me, the program gets a bit more practical, but not enough. There's a practicum experience at the end of the program, but I need more hands-on work now. So I've started working at a halfway house for people discharged from mental hospitals. But that's not working out the way I expected either. There's something about this whole helping business that is making me think twice about myself and about the profession."

A key blind spot translated into a future-oriented new perspective.

_____

_____

A range of possibilities for a better future.

_____

_____

_____

3. A 44-year-old nun, a member of a counselor training group, has been talking about her dissatisfaction with her present job. Although a nurse, she is presently teaching in a primary school because, she says, of the "urgent needs" of that school. When pressed, she refers briefly to a history of job dissatisfaction. In

the group she has shown herself to be an active, intelligent, and caring woman who tends to speak and act in self-effacing ways. She mentions how obedience has been stressed throughout her years in the religious order. She does mention, however, that things have been "letting up a bit" in recent years. The younger sisters don't seem to be as preoccupied with obedience as she is. She says: "The reason I'm talking about my job is that I don't want to become a counselor and then discover it's another job I'm dissatisfied with. It would be unfair to the people I'd be working with and unfair to my religious order, which is paying for my education. Of course, I have no iron-clad assurance that I'll be put in a job that will enable me to use my counselor training."

A key blind spot translated into a future-oriented new perspective.

_____

_____

A range of possibilities for a better future.

_____

_____

_____

_____

4. Marcus is a successful, 47-year-old businessman whose life is in fairly good order. However, Yves, his younger brother, is struggling financially. His credit-card debt is enormous. For Marcus, the problem is extending a helping hand to his younger brother without coming off as pitying, or being seen that way. He explains his dilemma to a colleague at work, "Yves is very sensitive to this older brother-younger brother stuff. Anyway, he has it in his head that I see him as incompetent and any help I offer is seen through that filter. I suppose that I'll always be the successful big brother and he will always be the inept little brother. It's not that way really but.... What I can't do is let our past history force me to stand off to the side with my hands in my pockets and do nothing when I know I could be of some help. I'm thinking of just writing him a check and saying, "Let's just make a new beginning in every way."

A key blind spot translated into a future-oriented new perspective.

_____

_____

A range of possibilities for a better future.

_____

_____

_____

_____

_____

5. A man, now 64 years old, retired early from work when he was 62. He and his wife wanted to take full advantage of the "golden" years. But his wife died a year after he retired. At the urging of friends, he has finally come to a counselor. He has been exploring some of the problems his retirement has created for him. His two married sons live with their families in other cities. In the counseling sessions he has been dealing somewhat repetitiously with the theme of loss. He says: "I seldom see the kids. I enjoy them and their families a lot when they do come. I get along real well with their wives. But now that my wife is gone. . . (pause) . . . and since I've stopped working . . . (pause) . . . I seem to just ramble around the house aimlessly, which is not like me at all. I suppose I should get rid of the house, but it's filled with a lot of memories – bittersweet memories now. There were a lot of good years here."

A key blind spot translated into a future-oriented new perspective.

_____

_____

A range of possibilities for a better future.

_____

_____

_____

_____

EXERCISE 11-3: HELPING SOMEONE DEVELOP POSSIBILITIES FOR A BETTER FUTURE

In this exercise, you are asked to help one of the other members of your group develop possibilities for a better future.

1. Divide up into groups of three for this exercise.
2. Assume, sequentially, the roles of client, helper, and observer.
3. In the role of client, provide a summary of one of the problem situations or undeveloped opportunities you focused on in any of the previous exercises. Or, if you prefer, choose a new issue.
4. In the role of counselor, listen to your partner summarize their problem situation or undeveloped opportunity. Use empathy and probes to get a full statement of the issue.
5. Finally, help the other develop possibilities for a better future. Use empathy, probes, summaries, and invitations to self-challenge to help the other expand the list.

Example: Geraldo, a junior in college majoring in business studies, has given the following summary of the problem situation: "My uncle runs a small business. He's offered me a kind of internship in the family business. While such an opportunity fits perfectly with my career plans, I have done nothing to develop it. Right now, it's only an interesting idea. But it still seems too good an opportunity to miss." Trish, his helper, uses all the communication skills learned up to this point to help him develop possibilities for a productive internship. Some of the dialogue goes like this:

TRISH: "So you haven't taken the time yet to spell out what the internship could look like."

GERALDO: "No. Generally, I see the internship as something that would bring to life the stuff I'm getting out of business books right now."

TRISH: "So the internship would give a chance for some hands-on experience with business realities that are just concepts now. What could the internship look like?"

GERALDO: "Well, I don't know what's on my uncle's mind. Though he's not going to push me into anything I don't want."

TRISH: "Then it seems to make sense for you to have some ideas in mind before you talk with him about it. What would *you* like the internship to look like? What do you want?"

GERALDO: "Let's see. I'd be putting in 10 to 15 hours a week at my uncle's business. Actually, that would make sense since I'm more or less overbalanced right now on social activities."

TRISH: "What would you be doing during those hours?"

GERALDO: "Well, I don't like the finance part of my program. So one thing I could do is become familiar with the finance part of his business—where the money comes from, what kind of debt he carries, cash flow—all those things that are still too theoretical for me in the courses I'm taking."

TRISH: "You'd see finance in action. It might even seem more interesting to you then. What else do you picture yourself doing?"

GERALDO: "I like to find out what management and supervision actually looks like inside a company. I get a confused picture of both of these from what I'm reading."

TRISH: "So it's one thing to read management books, but another to see how it actually works. What might you do as a manager or supervisor?"

GERALDO: "I'm not sure he'd let me run anything important. But maybe I could run a project or two. And the project might involve working with others—you know, like a small team. I think I'd like to see how teamwork can affect the business. I guess what I'm saying is that my uncle's business could be a lab for me. I'd be contributing, but also I'd be learning a lot."

TRISH: "So project work and teamwork could be important parts of the lab. But it also sounds like you want to be a contributor in all this."

GERALDO. "Yeah. I'd like to learn, but I'd like to make a difference."

TRISH. "What kind of difference?"

The dialogue goes on in that vein. Geraldo, with the help of Trish, develops a lot of possibilities for the internship.

1. With your partner critique this dialogue. What did Trish do well? What might have been done better?
2. Then move on to your own dialogue.
3. After eight or ten minutes, get feedback from the client first, followed by the observer, as to how useful you have been in helping the client develop possibilities.
4. Repeat the process until each person has played each role.

TASK 2:
HELP CLIENTS MOVE FROM POSSIBILITIES TO CHOICES
"What Solutions Are Best for Me?"

Once clients have brainstormed possibilities for a better future, they need to make some choices. Answers to the questions, "What do you want?" and "What do you need?" are possible goals. Answers to the question, "What do you *really* need and want?" are the actual goals that constitute the change agenda. It is as if the client were to say, "Here is what I want in place of what I have right now." Goals are a critical part of an agenda for constructive change. For instance, Geraldo, the college student mentioned above, after brainstorming a number of possibilities for structuring his internship in his uncle's company, chooses four of the possibilities and makes these his agenda. He shares them with his uncle and, after some negotiation, comes up with a revised package they can both live with. Review Chapter 11, Task 2 in the text before doing these exercises.

EXERCISE 11-4: TURNING POSSIBILITIES INTO GOALS – A CASE

In this exercise, you are asked to do what Geraldo had to do—choose goals from among the possibilities. Possibilities should be chosen because they will best help you manage your problem situation or develop some opportunity. First, read the following case.

The Case. Vanessa, 46 years old, has been divorced for about a year. She has done little to restructure her life and is still in the doldrums. At the urging of a friend, she sees a counselor. With the counselor's help, she brainstorms a range of possibilities for a better future around the theme of "my new life as a single person." She is currently a salesperson in the women's apparel department of a moderately upscale store. She lives in the house that was part of the divorce settlement. She has no children. When she finally decided that she wanted children, it was too late. There was some discussion with her ex-husband about adopting a child, but it didn't get very far. The marriage was already disintegrating. Her visits to the counselor have reawakened a desire to take charge of her life and not just let it happen. She let her marriage happen and it fell apart. She is not filled with anger at her former husband. If anything, she's a bit too down on herself. Her grieving is filled with self-recrimination.

The helper asks her, "What do you want now that the divorce is final? What would you like your post-divorce life to look like? What would you want to see in place?" Vanessa brainstorms the following possibilities:

- A job in some area of the fashion industry.
- A small condo that will not need much maintenance on my part, instead of the house.
- The elimination of poor-me attitudes as part of a much more creative outlook on self and life.
- The elimination of waiting around for things to happen.
- The development of a social life. For the time being, a range of friends rather than potential husbands. A couple of good women friends.

138

- Getting into physical shape.
- A hobby or avocation that I could get lost in. Something with substance.
- Some sort of volunteer work. With children, if possible.
- Some religion-related activities, not necessarily established-church related. Something that deals with the "bigger" questions of life.
- An end to all the self-recrimination over the divorce.
- Resetting my relationship with my mother (who strongly disapproved of the divorce).
- Getting over a deep-seated fear that my life is going to be bland, if not actually bleak.
- Possibly some involvement with politics.

1. Read the list from Vanessa's point of view (even though you hardly know her).
2. Because Vanessa cannot possibly do all of these at once, she has to make some choices. Put yourself in her place. Which items would you include in your agenda if you were Vanessa? In what way would the agenda items, if accomplished, help manage the overall problem situation and develop opportunities?
3. Choose three and give your reasons for including each.
4. Share your package together with the reasons for each item in it with a learning partner.
5. Discuss the differences in the items chosen and the reasons for the choices. Note that there is no one right package.

My first choice.

_____

My reason for the choice

_____

_____

_____

My second choice.

_____

My reason for the choice

_____

_____

_____

My third choice.

_____

My reason for the choice

_____

_____

_____

EXERCISE 11-5: TURNING PERSONAL POSSIBILITIES INTO PERSONAL GOALS

In this exercise you are asked to review EXERCISE 11-1: PROBLEMS AND OPPORTUNITIES IN THE SOCIAL SETTINGS OF LIFE in which you brainstormed a range of possibilities for a better future for yourself.

1. Pick two of the three social-settings you explored.
2. In each, after reviewing the possibilities that you brainstormed, come up with a goal and your reason for it.
3. Use the criteria outlined in Chapter 11 in order to "fashion" or "design" the goals.
After completing the exercise, share and critique your goals with a learning partner.

1. First social setting.

_____

Goal for this setting.

_____

_____

Reason for the goal

_____

_____

_____

2. Second social-setting.

_____

Goal for this setting.

_____

_____

Reason for the goal

_____

_____

_____

EXERCISE 11-6: HELPING CLIENTS TURN POSSIBILITIES INTO GOALS

In this exercise you are asked to review EXERCISE 11-2: BRAINSTORMING POSSIBILITIES FOR A BETTER FUTURE—CASES, in which you brainstormed a range of possibilities for a better future for three different clients.

1. Pick two of the three cases in that exercise.
2. In each, turn the possibilities that you brainstormed into a workable goal or set of goals.
3. Use the criteria outlined in Chapter 11 to "fashion" or "design" the goals.
4. After completing the exercise, share and critique your goals with a learning partner.

1. First social-setting situation.

_____

_____

Possibilities turned into a goal.

_____

_____

_____

_____

_____

Reason for the goal.

_____

_____

2. Second social-setting situation.

_____

_____

Possibilities turned into a goal.

_____

_____

_____

_____

_____

_____

Reason for the goal.

_____

_____

EXERCISE 11-7: SHAPING GOALS—FROM BROAD TO SPECIFIC

Many clients set goals that are too vague to be accomplished. "I'm going to clean up my act" is nothing more than a good intention. It is not yet a goal. Part of shaping, then, is making goals specific enough to drive action. This exercise involves moving from vagueness to concreteness.

1. First, using the example below, evaluate the goal Tom and Carol have set for themselves in terms of its specificity. If you don't think it is specific enough, shape it, and make it more specific.
2. Then, in the cases that follow, get inside the client's mind and try to determine what he or she might need or want.
3. Finally, move from a vaguely stated good intention, to a broad aim, and then to a specific goal.
4. Finally, learning partners should share the goals they have written and provide feedback for each other. When necessary, goals should be reformulated.

Example: Forty-two year-old Tom and his 39-year-old wife, Carol, have been talking to a counselor about how poorly they relate to each other. They have explored their own behavior in concrete ways, have developed a variety of new perspectives on themselves as both individuals and as a couple, and now want to do something about what they have learned. In broad terms, they have agreed to stop blaming each other for their problems and take mutual responsibility for creating a better relationship. What follows is their first shot at moving from a vague to a more specific goal in the area of communication:

> **Good Intention**: "We've got to do something about the way we communicate with each other. It's just not healthy."
> **General Aim**: "We are constantly at each other over inconsequential issues and then it escalates. We'd like to do something about the way we bicker."
> **Specific Goal**: "We want to call a moratorium on what has become almost childish bickering. We want to stop second-guessing, picking on, correcting, one-upping, putting down, questioning, and being snide with each other. But we don't want to leave a vacuum. We need something positive to take its place. For instance, when we kid around, each of us should make himself or herself, not the other, the butt of any joke. When we do disagree about something important, we should avoid

overly personal heated arguments and instead engage in mutual problem solving based on the skilled helping process we are using in counseling."

First, critique the specific goals in the above example and then work on the cases below.

1. A 53-year-old woman is talking to her lawyer-counselor. She has recently received a significant sum of money from a deceased relative. "I suppose I knew my aunt was going to remember me, but I had no idea that it would be this much. I've spent most of my life managing small amounts of money over short periods of time—like until the end of the month. I have no experience with this kind of money. I could go to school. I could travel. I certainly will want to set aside some money for others in the family. I'm excited and stunned at the same time. I'm the deer in the headlights everyone talks about."

Statement of good intention.

_____

_____

Broad aim.

_____

_____

Specific goal.

_____

_____

_____

_____

_____

2. Troy, 30 years old, has been discussing the stress he has been experiencing during this transitional year of his life. Part of the stress relates to his job. He has been working as an accountant with a large firm for the past five years. He makes a decent salary, but he is more and more dissatisfied with the kind of work he is doing. He finds accounting predictable and boring. He doesn't feel that there's much chance for advancement in this company. Many of his associates are much more ambitious and much more inventive than he is.

Statement of good intention.

_____

_____

Broad aim.

_____

_____

Specific goal.

_____

_____

_____

_____

_____

3. A high-school senior is talking to a school counselor about college, and what kinds of courses she might take there. However, she also mentions, somewhat tentatively, her disappointment in not being chosen as valedictorian of her class. She and almost everyone else had expected her to be chosen. She says: "I know that I would have liked to have been the class valedictorian, but I'm not so sure that you are supposed to count on anything like that. They chose Jane. She'll be good. She speaks well and she's very popular. But no one has a right to be valedictorian. I'd be kidding myself if I thought differently. I've done better in school than Jane, but I'm not as outgoing or popular."

Statement of good intention.

_____

_____

Broad aim.

_____

_____

Specific goal.

_____

_____

_____

_____

144

4. Len, married with three teenage children, lost all the family's savings in a gambling spree. Under a lot of pressure from both his family and his boss, he started attending Gamblers Anonymous meetings. He seemed to recover, and he stopped betting on horses and ball games. A couple of years went by, and Len stopped going to the GA meetings because he "no longer needed them." He even got a better job and was recovering financially quite well. But his wife began to notice that he was on the phone a great deal with his broker. She confronted him and he reluctantly agreed to a session with one of the friends he made at GA. He says, "She's worried that I'm gambling again. You know, it's really just the opposite. When I was gambling I was financially irresponsible. I lost our future on horses and ball games. But now I'm taking a very active part in creating our financial future. Every financial planner will tell you that investing in the market is central to sound financial planning. I'm a doer. I'm taking a very active role."

Statement of good intention.

_____

_____

Broad aim.

_____

_____

Specific goal.

_____

_____

_____

_____

5. Joan, a 32-year-old married woman, has two small children. Her husband has left her and she has no idea where he is. She has no relatives in the city and only a few acquaintances. She is talking to a counselor in a local community center about her plight. Since her husband was the breadwinner, she now has no income and no savings on which to draw.

Statement of good intention.

_____

_____

Broad aim.

_____

_____

145

Specific goal.

_____

_____

_____

_____

EXERCISE 11-8: SHAPING GOALS—MAKING SURE THAT THEY ARE WORKABLE

Accomplishing goals takes a lot of work. If the goals chosen are themselves flawed, then the job of accomplishing them becomes almost impossible. For instance, if Radcliffe chooses a career for which he does not have the talent, he is doing himself no favor. Hard work won't make up for the talent deficit. In order to be workable, goals generally need to be:

- stated as outcomes,
- specific enough to drive action,
- substantial enough to make a difference,
- characterized by the right mixture of risk and prudence,
- realistic,
- sustainable,
- flexible,
- congruent with the client's values,
- set in a reasonable time frame.

These characteristics can be seen as tools that help you shape both your own and clients' goals.

1. With a learning partner return to the four goals that each of you set in the previous two exercises. Use the above characteristics to review them once more.
2. What changes might you make in each of these goals to make them conform to the above characteristics?

EXERCISE 11-9: HELPING LEARNING PARTNERS SET WORKABLE GOALS

In this exercise you are asked to act as a helper to a learning partner. This exercise is done verbally.

1. The total training group is to be divided up into smaller groups of three.
2. There are three roles: client, helper, and observer. Decide the order in which you will play each role.
3. The client summarizes some problem situation, and then declares their intent to do something about the problem or some part of it.
4. The group member (in helper role), using empathic highlights, probing, and challenging, helps the client move from this statement of intent to a broad aim and then to a specific problem-managing or opportunity-developing goal that has the characteristics listed above.
5. When the helper feels that he or she has fulfilled this task, the session is ended and both observer and client give feedback to the helper on his or her effectiveness.
6. Repeat the process until each person has played each role.

Example: Since most students do not operate at 100% efficiency, there is usually room for improvement in the area of learning. Luisa, a junior beginning her third year of college, is dissatisfied with the way she goes about learning. She decides to use her imagination to invent a more creative approach to study. She brainstorms possibilities for a better study future, that is, goals that would constitute her new learning style. She says to herself: "In my role as student or learner, what do I want? Let me brainstorm the possibilities."

- I will stop studying for grades and start studying to learn. Paradoxically this might help my grades, but I will not be putting in extra effort just to raise a B to an A.
- I will be a better contributor in class, but not just in the sense that I will be trying to make a good impression on my teachers. I will do whatever I need to do to learn. This may mean placing more demands on teachers to clarify points, making more contributions, and involving myself in discussions with peers.
- I will be reading more broadly in the area of my major, psychology, not just the articles and books assigned but also in the areas of my interest. I will let my desire to know drive my learning.

She is particularly disappointed in the way she goes about writing the papers assigned in class. She believes that she wastes lots of time and ends up frustrated, so she picks this topic for the exercise.

> **Declaration of intent**: "I'm going to finally do something about the way I write papers for class."
>
> **Overall aim**: "I'm going to put in place a systematic but flexible process for writing papers for class."

The task of the training group helper is to help Luisa develop and shape a problem-managing goal. Luisa begins by sharing her declaration of intent and overall aim. Here is some of the dialogue that follows:

JEREMY: "What bothers you most about the way you have been doing your papers?"

LUISA: "Everything piles up at the end, and as a result the papers are rushed and practically never represent the kind of work I'm capable of."

JEREMY: "So the unsystematic approach drives out quality. Let's see what a more systematic approach might look like. What are some of the elements of the process you want to put in place?"

LUISA: "Well, I've thought about this a bit. For instance, once a paper is assigned, I will start a file on the topic and collect ideas, quotes, and data as I go along."

JEREMY: "So you'll start early and have an organized filing system."

LUISA: "Yes. I'll start with a single folder and just throw in ideas or quotes from books or parts of lecture notes. Ideas from anywhere or everywhere. I'll sort them out later."

JEREMY: "It sounds like the sorting process could become a bottleneck."

LUISA: "Hmm. You mean I'll end up with wads of stuff and wonder where to start. Well, there are two things I can do. I'll put only one item to a sheet of paper. One idea. One quote. You know. Then after two or three weeks I'll read through and separate them into different categories and organize them in folders."

147

JEREMY: "The sub-category approach."

LUISA: "Right. Later on I can decide which topics are the most important ones. Then I can . . . ."

Luisa and Jeremy go on in this vein until she has a clear idea of the process that she wants to put in place. When Luisa feels that she has a goal that will drive action, they stop, review, and fine-tune the goal in terms of the goal characteristics outlined above. Prepare for your session with your helper by thinking of a statement of good intention, and then by translating it into a broad goal. This is followed by a dialogue with a partner through which, with his or her help, you translate a broad intention into a workable goal.

Statement of good intention.

_____

_____

Broad aim.

_____

_____

First, share your good intention and broad aim with your partner. Then have a goal-setting dialogue based on the criteria for effective goals. When finished, change roles.

EXERCISE 11-10: RELATING GOAL CHOICE TO ACTION

Once clients state what they want, they need to move into action in order to get what they want. At this point there are two ways of looking at the relationship of goals to action, one formal and one informal.

> **Formal Action**. Stage III of the helping model—brainstorming action strategies, choosing the best package, and turning them into a plan for accomplishing goals—is the formal approach to action.
> **Informal Action**. Once clients get a fairly clear idea of what they want, there is no reason why they cannot move immediately into action, and do *something* that will move them in the direction of their goals. These are the "little" actions that can precede the formal planning process. Indeed these little actions can help clarify and fine-tune goals. This exercise is about these little actions.

1. State three personal goals that have emerged from your work in the training group.
2. For each goal list some "little actions" in which you might be able to engage in order to begin moving quickly in the direction of the goal.
3. Get feedback from a learning partner on both the goals and the actions.

Example. Review the case of Vanessa discussed above in Exercise 11-4. While she believes that it is essential for her to develop a better attitude about herself and life in general, she believes that it is not best to try to do this directly. Therefore, she chooses goals which will have a better attitude as a by-product.

> One goal centers on the development of a social life. For the time being, she wants a range of friends rather than an intimate partner or potential spouse. She would especially like a couple of good women friends.

A second goal revolves around religion in some sense of that term. She would like to have religion-related activities, but not necessarily church-related, as part of her life. Something that deals with the "bigger" questions of life.

A third goal relates to physical well-being. The stress of the divorce has left her exhausted. She wants to build herself up physically and get into good physical shape.

Vanessa looks at the third goal and has this to say "There are some things I can do immediately without coming up with a formal physical fitness program. First of all exercise. I can begin by walking—to work, to the store, and so on. Second, I can easily cut down a lot on fast food and add some fresh fruit and vegetables to my diet. Let me try these two things and see where it leads." Now move on to your own personal goals.

First personal goal.

_____

_____

Actions you can take immediately.

_____

_____

_____

_____

Second personal goal.

_____

_____

Actions you can take immediately.

_____

_____

_____

_____

_____

149

Third personal goal.

_____

_____

Actions you can take immediately.

_____

_____

_____

_____

_____

TASK 3:
HELP CLIENTS COMMITMENT THEMSELVES TO THEIR GOALS
"What Am I Willing to Pay for What I Want?"

Many of us choose goals that will help us manage problems and develop opportunities, but we do not explore the level of commitment that will be needed to pursue these goals successfully. Just because goals are tied nicely to the original problem situation, and initially are espoused by us, does not mean that we will follow through. Even goals which seem attractive from a cost/benefit point of view can fall by the wayside. The costs of accomplishing goals might not seem very high until it is time to pay up. Therefore, helping clients take a good look at commitment can raise the probability that they will actually pursue and accomplish these goals. Read the section on Task 3 in Chapter 11 before doing these exercises.

EXERCISE 11-11: MANAGING YOUR COMMITMENT TO YOUR GOALS

In this exercise you are asked to review the goals you have chosen to manage some problem situation or pursue some opportunity. The review should focus on commitment. It is not a question of challenging your good will. We all, at one time or another, make commitments that are not right for us.

1. Review the problem situations and unexplored opportunities you have been examining in this training program, together with the goals you have established for yourself.
2. Choose two important goals you have decided to pursue.
3. Read the following questions and answer the ones you see relevant to each of your goals, in order to gauge your level of commitment:

- What is your state of readiness for change in this area at this time?
- To what degree are you choosing this goal freely?
- To what degree are you choosing this goal from among a number of possibilities?
- How appealing is this goal to you?
- Name any ways in which your goal does not appeal to you.
- What's driving you to choose this goal?

150

- If your goal is in any way being imposed by others, what are you doing to make it your own? What incentives are there besides mere compliance?
- What difficulties are you experiencing in committing yourself to this goal?
- To what degree is it possible that your commitment is not a true commitment?
- What can you do to get rid of the disincentives and overcome the obstacles?
- What can you do to increase your commitment?
- In what ways can the goal be reformulated to make it more appealing?
- To what degree is the timing for pursuing this goal right or wrong?

Evaluation of your commitment to your first goal.

_____

_____

_____

_____

_____

Evaluation of your commitment to your second goal

_____

_____

_____

_____

_____

4. With a learning partner, review your principal learnings from answering the above questions about these two goals. Use empathic highlights, probes, and challenges to help one another explore levels of commitment.

5. Finally, to the degree necessary, reformulate both goals in terms of what you have learned from the dialogue.

Restated goal #1.

_____

_____

_____

_____

_____

Restated goal #2.

_____

_____

_____

_____

_____

EXERCISE 11-12: REVIEWING BENEFITS VERSUS COSTS IN CHOOSING GOALS

In most choices we make there are both benefits and costs. Commitment to a goal often depends on a favorable cost/benefit ratio. To what extent do the benefits outweigh the costs? The balance-sheet methodology, outlined in Chapter 12 of *The Skilled Helper*, can be useful in determining whether goals being chosen offer "value for money," as it were. Consider the following case.

Case. One January, Helga, a married woman with two children, one a senior in college and one a sophomore, was told that she had an advanced case of cancer. She was also told that a rather rigorous series of chemotherapy treatments might prolong her life, but they would not save her. She desperately wanted to see her daughter graduate from college in June, so she opted for the treatments. Although she found them quite difficult, she buoyed herself up by the desire to be at the graduation. Although in a wheel chair, she was there for the graduation in June. When the doctor suggested that she could now face the inevitable with equanimity, she said: "But, doctor, in only two years my son will be graduating."

This is a striking example of a woman's deciding that the costs, however high, were outweighed by the benefits. Obviously, this is not always the case. Benefits include incentives and rewards for whatever source, while costs include time and effort spent, resources used, psychological wear and tear, battling negative emotions, and the like. This exercise gives you the opportunity to explore your goals from a cost-benefit perspective. To what degree is it worth the effort? What's the payoff?

1. Divide up into pairs, with one partner acting as client, one as helper.
2. Help your partner review one of the goals of their agenda from a cost/benefit perspective. The helper is to use empathic highlights, probing, and challenging to help his or her partner do this. Help your partner identify benefits and costs, and do some kind of trade-off analysis, such as the balance-sheet technique.
3. Help you partner clearly state the incentives and payoffs that enable her or him commit to the specified goals.

Example: Carlita, a 28-year-old and a third-year doctoral student in clinical psychology, is discussing her commitment to her career goals with Eban, a fellow student. She is an only child and both parents and grandparents have been pressing her to get married and have children. She wants to establish herself in her career first. She has one more year to go in the program plus a year's internship.

CARLITA: "If I don't get my degree and begin my career before getting married, I'll never finish. I know others can do this. But I would be feeling all sorts of pressure from family and friends."

EBAN: "So, from your perspective, it's almost now or never. But you're already under a lot of pressure. What about the cost-benefit equation?"

CARLITA: "I'm the first woman to break the mold in my extended family. So I have to expect some costs. I don't like hearing disappointment in the voices of my parents and grandparents. But I'm setting a direction for the rest of my life."

EBAN: "You're ready to pay the price. How do you counter the pressures?"

CARLITA: "First of all, I don't avoid them. I talk to them about how exciting I find my work. I talk to them about how this will benefit both me and whatever family I have in the future. Without saying it in these words, I ask them to be proud of me, not disappointed. I'm still the same daughter and granddaughter that they have always loved. I let them know by my actions that I love them. If I were to give in, that wouldn't be love. In my own way I let them know how *I* need to be loved."

EBAN: "What about social life? Any pressures there?"

CARLITA: "I have a couple of close men friends. But they know I'm not going to get into a very serious relationship right now. Eventually, I want someone who takes me as I am."

EBAN: "So there are a few bumps there, too."

CARLITA: "None that I can't handle. I see myself as responsible and focused. If they want to see that as bullheadedness, well, that's their problem. Furthermore . . . "

And the dialogue continues in this vein.

4. After ten minutes of discussion, stop and receive feedback from the client and the person being helped.
5. After the discussion, each member of the training group should get a new partner, change roles, and repeat the process.

EXERCISE 11-13: COMPETING AGENDAS

When setting goals, clients often fail to see them in the context of the current range of demands on their time, energy, and resources. It is no wonder, then, that some goals do not get accomplished because they are squeezed out by all the other things clients have to do. In other words, clients have to face up to the reality of "competing agendas." Otherwise, they end up trying to carry impossible burdens. Therefore, competing agendas need to be reviewed lest they block or dilute goals set during the helping process.

Example. Ivor, 24 years old, is talking to members of his training group about a his lack of progress in learning both the helping model and the skills that make it work. Now, believing he knows what is happening, he says: "I came to this program all fired up to become the best counselor possible. But it hasn't quite happened. I know I'm not lazy. But this past week it struck me. I really enjoy the social life of this group. Lots of us have become friends. So much so, that my social life with the group has been doing two things. First, I'm letting it take up a lot of my time. I put off the work I should be doing.

Second, I'm reluctant to challenge my friends in both group interactions and in one-to-one practice. This is being unfair to myself and unfair to you. I certainly want to enjoy the social life here, but I don't want it to compete with my original goal."

1. Look into two different goals you have set for yourself over the course of this training program.
2. Review the current commitments of your life both within the training group and outside. Outline the agendas that currently take up a great deal of your time, energy, and interest.
3. Indicate any competing agendas that might prevent you from accomplishing your goal.
4. Outline what needs to be done to manage competing agendas. This could mean dropping some current activities or reducing the scope of the goal.

Competing agendas for first goal.

_____

_____

_____

_____

_____

Ways of managing competing agendas.

_____

_____

_____

_____

_____

Competing agendas for second goal.

_____

_____

_____

_____

Ways of managing competing agendas.

_____

_____

_____

_____

_____

Chapter 12

STAGE III: THE WAY FORWARD—
HELP CLIENTS DEVELOP PLANS TO ACCOMPLISH GOALS

Stage III is an important part of solution-focused phase of helping. Once clients are helped to establish goals, they often still need help to plan the actions or strategies that will enable them to accomplish their goals. These actions or strategies are solutions with a small 's'. Stage III has three tasks all aimed at helping clients move to goal-accomplishing action.

Task 1: Possible Strategies. Help clients brainstorm strategies for accomplishing their goals. "What kind of actions will help me get what I need and want?"
Task 2: Best-Fit Strategies. Help clients choose strategies tailored to their preferences and resources. "What set of actions are best for me?"
Task 3: Plans. Help clients formulate actionable plans. "What should my campaign for constructive change look like? What do I need to do first? Second? Third?"

Stage III adds the final pieces to a client's planning a program for constructive change. Stage III deals with the "game plan." However, these three "tasks" constitute *planning* for action and should not be confused with action itself. Without action, a program for constructive change is nothing more than a wish list. The formal implementation of plans is discussed in Chapter 13.

TASK 1:
HELP CLIENTS BRAINSTORM STRATEGIES
FOR ACCOMPLISHING THEIR GOALS:

There is usually more than one way to accomplish a goal. However, clients often focus on a single strategy or just a few. The task of the counselor in Task 1 is to help clients discover a number of different routes to goal accomplishment. Clients tend to choose a better strategy or set of strategies if they choose from among a number of possibilities. Read Chapter 12 before doing the exercises in this section.

EXERCISE 12-1: BRAINSTORMING ACTION STRATEGIES FOR YOUR OWN GOALS

Brainstorming is a technique you can use to help yourself and your clients move from constricted to imaginative or inventive thinking. Recall the rules of brainstorming but remember this one caution: Remain *focused* on the issue at hand; let your imagination roam freely, but stay within the broad confines of the problem situation at hand.

- Encourage focused quantity. Deal with the quality of suggestions later.
- Do not criticize any suggestions. Merely record them.
- Combine suggestions to make new ones.
- Encourage even wild possibilities that stay focused on the problem situation. For instance, "One way to keep to my diet and lose weight is to have my mouth or some part of my eating apparatus sewn up."
- When you feel you have said all you can say, put the list aside and come back to it later and add fresh ideas, but remain focused.

Example: Ira, a retired lawyer who is training to be a counselor, is in a high-risk category for a heart attack: some of his relatives have died relatively early in life from heart attacks, he is overweight, he exercises very little, he is under a great deal of pressure in his job, and he smokes over a pack of cigarettes a day. One of his goals is to stop smoking within a month. With the help of a nurse practitioner friend, he comes up with the following list of strategies:

- just stop cold turkey.
- shame myself into it, "How can I be a helper if I engage in self-destructive practices such as smoking?"
- cut down, one less per day until zero is reached.
- use nicotine-based aids, such as patches and gum.
- look at movies of people dying with lung cancer.
- pray for help from God to quit.
- use those progressive filters which gradually squeeze all taste from cigarettes.
- switch to a brand that doesn't taste good.
- switch to a brand that is so heavy in tars and nicotine that even I see it as too much.
- smoke constantly until I can't stand it any more.
- let people know that I'm quitting.
- put an ad in the paper in which I commit myself to stopping.
- send a dollar for each cigarette smoked to a cause I don't believe in, for instance, the "other" political party.
- get hypnotized; through a variety of post-hypnotic suggestions have the craving for smoking lessened.
- pair smoking with painful electric shocks.
- take a pledge before my minister to stop smoking.
- join a group for professionals who want to stop smoking.
- visit the hospital and talk to people dying of lung cancer.
- if I buy cigarettes and have one or two, throw the rest away as soon as I come to my senses.
- hire someone to follow me around and make fun of me whenever I have a cigarette.
- have my hands put in casts so I can't hold a cigarette.
- don't allow myself to watch television on the days in which I have even one cigarette.
- reward myself with a week-end fishing trip once I have not smoked for two weeks.
- void friends who smoke.
- have a ceremony in which I ritually burn whatever cigarettes I have and commit myself to living without them.
- suck on hard candy made with one of the non-sugar sweeteners instead of smoking.
- give myself points each time I want to smoke a cigarette and don't; when I have saved up a number of points reward myself with some kind of "luxury."

Note that Ira includes a number of wild but focused possibilities in his brainstorming session. Now follow these steps.

1. On separate piece of paper, brainstorm ways of achieving each of the goals you list below. Add wilder possibilities at the end.
2. After you have finished your list, take one of the goals and the brainstorming list, sit down with a learning partner, and through dialogue, expand your list. Follow the rules of brainstorming.
3. Switch roles. Through empathic highlights, probing, summarizing, and challenge, help your partner expand the list.
4. Keep both lists of brainstormed strategies. You will use them in a later exercise.

Goal # 1. _____

_____

Goal # 2. _____

_____

EXERCISE 12-2: ACTION STRATEGIES—PUTTING YOURSELF IN THE CLIENT'S SHOES

Here are a number of cases in which the client has formulated a goal and needs help in determining how to accomplish it. You are asked to put yourself in the client's shoes and brainstorm action strategies that you yourself might think of using were you that particular client.

Example: Renalda, 53 years old, has been a very active person—career-wise, physically, socially, and intellectually. In fact, she has always prided herself on the balance she has been able to maintain in her life. However, an auto accident that was not her fault has left her a paraplegic. With the help of a counselor, she has begun to manage the depression that almost inevitably follows such a tragedy. In the process of re-directing her life, she has set some goals. Since her job and her recreational activities involved a great deal of physical activity, a great deal of re-direction is called for.

One of her goals is to write a book called "The Book of Hope," about ordinary people who have creatively re-set their lives after some kind of tragedy. The book has two purposes. Since it would be partly autobiographical, it would be a kind of chronicle of her re-direction efforts. This will help her commit herself to some of the grueling rehabilitation work that is in store for her. Second, since the book would also be about others struggling with their own tragedies, these people will be models for her. Renalda has never published anything, so the "how" is more difficult. For her, writing the book is more important than publishing it. Therefore, the anxiety of finding a publisher is not part of the "how." Here are one person's set of brainstormed possibilities for the "if I were Renalda" case.

- Get a book on writing and learn the basics.
- Start writing short bits on my own experience, anything that comes to mind.
- Read books written by those who conquered some kind of tragedy.
- Talk to the authors of these books.
- Find out what the pitfalls of writing are like, for instance, writer's block.
- Get a ghostwriter who can translate my ideas into words.
- Write a number of very short, to-the-point pamphlets, and then turn them into a book.
- Learn how to use a word-processing program both as part of my physical rehabilitation program and as a way of jotting down and playing with ideas.
- Do rough drafts of topics that interest me and let someone else shape them into written stories.
- Record discussions about my own experiences with the counselor, the rehabilitation professionals, and friends and then have these transcribed for editing.
- Interview people who have turned tragedies like mine around.
- Interview professionals and the relatives and friends of people involved in personal tragedies. Record their points of view.
- Through discussion with friends get a clear idea of what this book will be about.
- Find some way of making it a bit different from similar books. What could I do that would give such a book a special slant?

Now do the same kind of work for each of the following cases.

Case # 1. Emma, 27 years old, has been a heroin addict. Her life has revolved around her habit. She has been told by her employer, an ad agency that, despite her creativity, her erratic performance at work will no longer be tolerated. She opts for a treatment – known as ultrarapid opiate detoxification – in which the drug naltrexone replaces heroin in the addict's opium receptors within six hours or so. Treatment with naltrexone is to continue for six months. Now that she is free of opiates, the challenge is to build a very different lifestyle – one that will focus her talents and energies. She is in opportunity-development mode. This includes a new social life since, currently, she has no social life to speak of. One of her goals is reconciliation with her family, including her parents, one sister, and two brothers, together with their families. She has seen very little of them over the past three years, with the exception of attending two weddings. She wants to start the work on reconciliation while she is still feeling well. She realizes that reconciliation is a two-way street, and that she cannot set goals for others.

Goal. If you were Emma, what would "reconciliation with my family" look like? If accomplished, what would this goal look like? What would be in place that is not now in place? How is the goal modified by the fact that reconciliation is a two-way street? Formulate a goal that has the characteristics of a viable goal outlined in Chapter 11.

_____

_____

_____

_____

Brainstorming Strategies. What are some of the things you might do to achieve the kind of reconciliation with your family that you have outlined above? Do the focused brainstorming on a separate sheet of paper. Remember to include some "wild," but still focused, possibilities.

1. After you have developed your list, first share your goal with a learning partner. Give each other feedback on the quality of the goal. Note especially how you and your partner interpret "reconciliation."
2. Next share the lists of brainstormed possibilities. Working together, add several more possibilities to the combined lists.
3. Discuss with your partner what you have learned from this exercise.

Case # 2. A priest was wrongfully accused of molesting a boy in his parish, at a time when everyone who was suspected was considered guilty. Working with a counselor, he set three goals. His "now" goal was to maintain his equilibrium under stress. Since, at the time, other priests had been accused and convicted of pedophilia, he knew that in the eyes of many he would be seen to be guilty until proven innocent. With the help of the counselor and some close friends, both lay and clerical, he kept his head above water. A near-term goal was to win the case in court. He also accomplished this goal. He was acquitted and all charges were dropped. After the trial, the bishop wanted to send him to a different parish "to start fresh." He had been removed from his pastorate and was living at the seminary. But he wanted to return to the same parish and re-establish his relationship with his parishioners. After all, he had done nothing wrong. The bishop agreed to reinstate him in his parish.

Goal. If you were this man, what would "re-establishing my relationship with reconciliation with my parishioners" look like? What would be in place that is not now in place? Formulate a goal that has the characteristics of a viable goal outlined earlier.

_____

_____

_____

_____

Brainstorming Strategies. What are some of the things you might do to re-establish your relationship with your parishioners, especially in view of the fact that some might still see you as tainted by the whole affair? Do the brainstorming on a separate sheet of paper. Include some "wild" possibilities.

Now complete steps 1-3 as above.

EXERCISE 12-3: HELPING OTHERS BRAINSTORM STRATEGIES FOR ACTION

As a counselor, you can help your fellow trainees stimulate their imaginations to come up with creative ways of achieving their goals. In this exercise, use probes and challenges based on questions such as the following:

> **How**: How can you get where you'd like to go? How many different ways are there to accomplish what you want to accomplish?
> **Who**: Who can help you achieve your goal? What people can serve as resources for the accomplishment of this goal?
> **What**: What resources both inside yourself and outside can help you accomplish your goals?
> **Where**: What places can help you achieve your goal?
> **When**: What times or what kind of timing can help you achieve your goal? What times might be more suitable?

Example: What follows are bits and pieces of a counseling session in which Arnold the counselor trainee, is helping Meredith, playing the role of the client, develop strategies to accomplish one of her goals. One of Meredith's problems is that she procrastinates a great deal. She feels that she needs to manage this problem in her own life if she is to help future clients move from inertia to action. Therefore, her "good intention" is to reduce the amount of procrastination in her life. In exploring this problem, she realizes that she puts off many of the assignments she receives in class. The result is that she is overloaded at the end of the semester, experiences a great deal of stress, does many of the tasks poorly, and receives lower grades than she is capable of. While her overall goal is to reduce the total amount of procrastination in her life, her immediate goal is to be up-to-date every week in all assignments for the counseling course. She also wants to finish the major paper for the course one full week before it is due. She chooses this course

160

as her target because she finds it the most interesting, and has many incentives for doing the work on time. She presents her list of the strategies she has brainstormed on her own to Angie. After discussing this list, their further conversations sound something like this:

ARNOLD: "You said that you waste a lot of time. Tell me more about that."

MEREDITH: "Well, I go to the library a lot to study. And I go with the best intentions But I meet friends, we kid around, and time slips away. I guess the library is not the best place to study."

ARNOLD: "Does that suggest another strategy?"

MEREDITH: "Yeah, study someplace where none of my friends is around. But then I might not—"

ARNOLD (interrupting): "We'll evaluate this later. Right now let's just add it to the list."

* * * * *

ARNOLD: "I know it's your job to manage your own problems, but I assume that you could get help from others and still stay in charge of yourself. Could anyone help you achieve your goal?"

MEREDITH: "I've been thinking about that. I have one friend. . . we make a bit of fun of him because he makes sure he gets everything done on time. He's not the smartest one of our group, but he gets good grades because he knows how to study. I'd like to pair up with him in some way, maybe even anticipate deadlines the way he does."

* * * * *

ARNOLD: "Your strategy list sounds good but a bit tame. Maybe it sounds wild to you because you're trying to change what you do."

MEREDITH: "I guess I could get wilder. Hmmm. I could make a contract with my counseling prof to get the written assignments in early! That would be wild for me."

Note here that Arnold uses probes based on the how, who, what, when, and where probes outlined above. he also uses the "wilder possibilities" probe.

1. Divide up into groups of three: client, helper, and observer.
2. Decide in which order you will play these roles.
3. The client will briefly summarize a concern or problem and a specific goal which, if accomplished, will help them manage the problem or develop the unused opportunity more fully. Make sure that the client states the goal in such a way that it fulfills the criteria for a viable goal.
4. Give the client five minutes to write down as many possible ways of accomplishing the goal as they can think of.
5. Then help the client expand the list. Use probes and challenges based on the questions listed above.
6. Encourage the client to follow the rules of brainstorming. For instance, do not let them criticize the strategies as they brainstorm, but help them stay focused on the problem situation.
7. At the end of the session, stop and receive feedback from your client and the observer as to the helpfulness of your probes and challenges.
8. Switch roles and repeat the process until each has played all three roles.

TASK 2:
HELP CLIENTS CHOOSE BEST-FIT STRATEGIES
"What Actions Are Best For Me?

The principle is simple. Strategies for action chosen from a large pool of strategies tend to be more effective than those chosen from a small pool. However, if brainstorming is successful, clients are sometimes left with more possibilities than they can handle. Therefore, once clients have been helped to brainstorm a range of strategies, they might also need help in choosing the most useful. These exercises are designed to help you help clients choose "best-fit" strategies, that is, strategies that best fit the resources, style, circumstances, and motivation level of clients.

EXERCISE 12-4: USING CRITERIA TO CHOOSE BEST-FIT STRATEGIES

Just as there are criteria for crafting goals (Step II-B), so there are criteria for choosing best-fit strategies. The following questions can be asked especially when the client is having difficulty choosing from among a number of possibilities. These criteria complement, rather than take the place of, common sense.

> **Clarity.** Is the strategy clear?
> **Relevance.** Is it relevant to my problem situation and goal?
> **Realism.** Is it realistic? Can I do it?
> **Appeal.** Does it appeal to me?
> **Values.** Is it consistent with my values?
> **Efficacy.** Is it effective enough? Does it have bite? Will it get me there?

Example: Ira, the counselor trainee who wanted to quit smoking, considered the following possibility on his list: "Cut down gradually, that is, every other day eliminate one cigarette from the 30 I smoke daily. In two months, I would be free."

> **Clarity**: "This strategy is very clear; I can actually see the number diminishing. It gets a 6 or 7 for clarity."
> **Relevance**: "It leads inevitably to the elimination of my smoking habit, but only if I stick with it."
> **Realism**: "I could probably bring this off. It would be like a game; that would keep me at it. But maybe too much like a game. Also I know others who have done it."
> **Appeal**: "Although I like the idea of easing into it, I'm quitting because I now am personally convinced that smoking is very dangerous for me. I should stop at once."
> **Values**: "There is something in me that says that I should be able to quit cold turkey. That has more moral appeal to me. For me there's something phony about gradually cutting down."
> **Effectiveness**: "The more I draw this action program out, the more likely am I to give it up. There are too many pitfalls spread out over a two-month period."

In summary, Ira says, "I now see that only strategies related to stopping cold turkey have bite. In fact, stopping is not the hard thing. Not taking smoking up again in the face of temptation, that's the real problem." The above criteria not only helped Ira eliminate all strategies related to a gradual reduction in smoking, but it helped him redefine his goal to "stopping and staying stopped." Sustainability is the real issue. He also noted that the brainstormed strategies he preferred referred not just to stopping but to sustainability. For instance, in the short term, the nicotine-flavored gum would reduce the craving once he has stopped.

162

1. Read the following case, put yourself in this woman's shoes, and, like Ira, use the criteria for determining the viability of the strategy she proposes.

Case: A young woman has been having disagreements with a male friend. Since he is not the kind of person she wants to marry, her goal is to establish a relationship with him that is less intimate-- one without sexual relations. She knows that she can be friends with him but is not sure if he can be just a friend with her. She would rather not lose him as a friend. She also knows that he sees other women. She uses the above criteria to evaluate the following strategy: "I'll call a moratorium on our relationship. I'll tell him that I don't want to see him for four months. After that we will be in a better position to re-establish a different kind of relationship, if that's what we want."

1. Once you have done your analysis, share your findings with a learning partner. See what the two of you can learn from your differences.
2. Next review the reasoning that the woman herself went through and her decision. Here, then, is her analysis:

> **Clarity**: "A moratorium is quite clear; it would mean stopping all communication for four months. It would be as if one of us were in Australia for four months. But no phone calls."
> **Relevance**: "Since my goal is moving into a different kind of relationship with him, stepping back to let old ties and behaviors die a bit is essential. A moratorium is not the same as ending a relationship. It leaves the door open. But it does indicate that cutting the relationship off completely could ultimately be the best course."
> **Realism**: "I can stop seeing him. I think I have the assertiveness to tell him exactly what I want and stick to my decision. Obviously I don't know how realistic he will think it is. He might see it as an easy way for me to brush him off. He might get angry and tell me to forget about it."
> **Appeal**: "The moratorium appeals to me. It will be a relief not having to manage my relationship with him for a while."
> **Values**: "There is something unilateral about this decision and I prefer to make decisions that affect another person collaboratively. On the other hand, I do not want to string him along, keeping his hopes up for a deeper relationship, or even marriage."
> **Effectiveness**: "When—and if, because it depends on him, too—we start seeing each other again, it will be much easier to determine whether any kind of meaningful relationship is possible. A moratorium will help determine things one way or another." Based on her analysis, she decides to propose the moratorium to her friend.

3. Discuss her reasoning with your learning partner. In what ways does your partner's analysis differ from your own? If you were her helper, how would you go about inviting her to challenge her reasoning and her decision?

EXERCISE 12-5: A PRELIMINARY SCAN OF BEST-FIT STRATEGIES FOR YOURSELF

You do not necessarily need sophisticated methodologies to come up with a package of strategies that will help you accomplish a goal. In this exercise you are asked to use your common sense to make a "first cut" on the strategies you brainstormed for yourself in Exercise 12-1.

1. Review the strategies you brainstormed for each of the goals in that exercise.
2. Star the strategies that make most sense to you. Just use common-sense judgment. Use the following example as a guideline.

Example: Let's return to the case of Ira in Exercise 12-2. Remember that he is the counselor trainee who wants to stop smoking. Here are the strategies he brainstormed. The ones he chooses in a preliminary common-sense scan are marked with an asterisk (*) instead of a bullet (•).

* just stop cold turkey.
* shame myself into it, "How can I be a helper if I engage in self-destructive practices such as smoking?"
• cut down, one less per day until zero is reached.
* use the new nicotine-based aids, like gum, advertised on TV.
• look at movies of people dying with lung cancer.
* pray for help from God to quit.
• use those progressive filters which gradually squeeze all taste from cigarettes.
• switch to a brand that doesn't taste good.
• switch to a brand that is so heavy in tars and nicotine that even I see it as too much.
• smoke constantly until I can't stand it any more.
• let people know that I'm quitting.
• put an ad in the paper in which I commit myself to stopping.
• send a dollar for each cigarette smoked to a cause I don't believe in, for instance, the "other" political party.
• get hypnotized; through a variety of post-hypnotic suggestions have the craving for smoking lessened.
• pair smoking with painful electric shocks.
• take a pledge before my minister to stop smoking.
• join a group for professionals who want to stop smoking.
• visit the hospital and talk to people dying of lung cancer.
• if I buy cigarettes and have one or two, throw the rest away as soon as I come to my senses.
• hire someone to follow me around and make fun of me whenever I have a cigarette.
• have my hands put in casts so I can't hold a cigarette.
• don't allow myself to watch television on the days in which I have even one cigarette.
* reward myself with a week-end fishing trip once I have not smoked for two weeks.
• have a ceremony in which I ritually burn whatever cigarettes I have and commit myself to living without them.
• suck on hard candy made with one of the non-sugar sweeteners instead of smoking.
* give myself points each time I want to smoke a cigarette and refrain from doing so; when I have saved up a number of points reward myself with some kind of "luxury."

3. Share with a learning partner the Exercise 12-1 strategies you have starred (with an asterisk) for yourself and the reasons for choosing them. What makes each of the starred strategies "best-fit" for you? Have a critical conversation with your learning partner about your decisions.

4. Switch roles. Have your partner share starred strategies and the reasons for them. Discuss the implications.

EXERCISE 12-6: BEST-FIT STRATEGIES—PUTTING YOURSELF IN THE CLIENT'S SHOES

In this exercise you are asked to put yourself in clients' shoes as they struggle to choose the strategies that will best enable them to accomplish their goals.

1. Review the case of Renalda in Exercise 12-3.
2. Review the strategies listed below that were brainstormed there.

- Get a book on writing and learn the basics.
- Start writing short bits on my own experience, anything that comes to mind.
- Read books written by those who conquered some kind of tragedy.
- Talk to the authors of these books.
- Find out what the pitfalls of writing are like, for instance, writer's block.
- Get a ghost writer who can translate my ideas into words.
- Write a number of very short, to-the-point pamphlets, and then turn them into a book.
- Learn how to use a word-processing program both as part of my physical rehabilitation program and as a way of jotting down and playing with ideas.
- Do rough drafts of topics that interest me and let someone else put them into shape.
- Record discussions about my own experiences with the counselor, the rehabilitation professionals, and friends and then have these transcribed for editing.
- Interview people who have turned tragedies like mine around.
- Interview professionals and the relatives and friends of people involved in personal tragedies. Record their points of view.
- Through discussion with friends get a clear idea of what this book will be about.
- Find some way of making it a bit different from similar books. What could I do that would give such a book a special slant?

3. If you have further strategies, add them to the list now.
4. Using your common sense, star the strategies you believe belong in the best-fit category.
5. Jot down the reasons for your choices. What makes them best-fit?
6. Share your choices and reasons with a learning partner and give each other feedback. Discuss what changes you would make to your choices.

EXERCISE 12-7: CHOOSING BEST-FIT STRATEGIES—THE BALANCE-SHEET METHOD

The balance sheet is another tool you can use to evaluate different program possibilities or courses of action. It is especially useful when the problem situation is serious, and you are having difficulty rating different courses of action.

Example: Rev. Alex M. has gone through several agonizing months re-evaluating his career in the ministry. He finally decides that he wants to leave the ministry and get an ordinary job. His decision, though painful in coming, leaves him with a great deal of peace. He now wonders just how to go about this. One possibility, now that he has made his decision, is to leave immediately. However, since this is a serious choice, he wants to look at it from all angles. He uses the balance sheet to evaluate the strategy of leaving his position at his present church immediately. We will not present his entire analysis (indeed, each bit of the balance sheet need not be used). Here are some of his key findings:

> **Benefits for me:** Now that I've made my decision to leave, it will be a relief to get away. I want to get away as quickly as possible.
>
> - *Acceptability*: I have a right to think of my personal needs. I've spent years putting the needs of others and of the institution ahead of my own. I 'm not saying that I regret this. Rather, this is now my season.
> - *Unacceptability*: Leaving right away seems somewhat impulsive to me, meeting my own needs to be rid of a burden.

Costs for me: I don't have a job and I have practically no savings. I'll be in financial crisis.

- *Acceptability*: My frustration is so high that I'm willing to take some financial chances. Besides, I'm well educated and the job market is good.
- *Unacceptability*: I will have to forego some of the little luxuries of life for a while, but that's not really unacceptable. I know I can build some kind of satisfying career.

Benefits for significant others: The associate minister of the parish would finally be out from under the burden of these last months. I have been hard to live with. My parents will actually feel better because they know I've been pretty unhappy.

- *Acceptability*: My best bet is that the associate minister will be so relieved that he will not mind the extra work. Anyway, he's much better than I at getting people involved in the work of the congregation.
- *Unacceptability*: I can't think of any particular downside here.

Costs to significant others: This is the hard part. Many of the things I do in this church are not part of programs that have been embedded in the structure of the church. They depend on me personally. If I leave immediately, many of these programs will falter and perhaps die because I have failed to develop leaders from among the members of the congregation. There will be no transition period. The congregation can't count on the associate minister taking over, since he and I have not worked that closely on any of the programs in question.

- *Acceptability*: The members of the congregation need to become more self-sufficient. They should work for what they get instead of counting so heavily on their ministers.
- *Unacceptability*: Since I have not worked at developing lay leaders, I feel some responsibility for doing something to see to it that the programs do not die. Some of my deeper feelings say that it isn't fair to pick up and run.

Alex goes on to use this process to help him make a decision. He finally decides to stay an extra six months and spend time with potential leaders within the congregation. He will tell them his intentions and then help them take ownership of essential programs.

1. Choose a personal goal for which you have brainstormed strategies.
2. Choose a major strategy or course of action you would like to explore much more fully. If this exercise is to be meaningful, the problem area, the goal, and the strategy or course of action in question must have a good deal of substance to them. Using the balance-sheet methodology for a trivial issue would be a waste of time.
3. Identify the "significant others" and the "significant social settings" that would be affected by your choice.
4. Explore the possible course of action by using as much of the balance sheet as is necessary to help you make a sound decision.
5. Share your balance sheet with a learning partner. Through dialogue, give one another feedback and help one another improve the choices.

TASK 3
HELP CLIENTS MAKE PLANS:
"What Kind of Plan Will Help Me Get What I Need And Want?"

A plan is a step-by-step procedure for accomplishing each goal of a change agenda. The strategies chosen in Task 2 often need to be organized and translated into a step-by-step plan. The plan should include only the kind of detail needed to drive action. Overly-detailed plans usually fall by the wayside. Clients are more likely to act if they know what they are going to do first, second, third, and so forth. Realistic time frames for each of the steps are also essential. The plan imposes the discipline clients need to get things done. To prepare for these exercises, read about Task 3 in Chapter 12 of the text.

EXERCISE 12-8: DEVELOPING YOUR OWN ACTION PLAN

In this exercise you are asked to establish a workable plan to accomplish one of the goals you have set for yourself.

1. Choose a problem situation or undeveloped opportunity that you have been working on, that is, one that you have explored and for which you have developed goals and action strategies.
2. Indicate the goal you want to accomplish.
3. Write a simple plan that embodies the principles outlined in Chapter 12.

Example. Emma, a 27-year-old heroin addict, who has just undergone an ultra-rapid opiate detoxification process and is now opiate-free, wants to establish or reestablish a social life that will include family and former friends. While on heroin, she either abandoned family and friends or they abandoned her. She has explored some ways of going about this with her counselor and settles on the following plan:

*   She will contact a couple of friends and a couple of members of the family who she thinks might be most receptive to her overtures. She will first write them and tell them what has happened and what she wants to do. She will follow up with a phone call.
*   Because she realizes that she might experience some rejection, she tells the members of her counseling group that she will need their support. Their support, both face-to-face and by phone, is an important part of her plan.
*   When she is sure of the help of both a friend and of a family member, she will arrange to meet them separately face-to-face. She will use these encounters to get the "lay of the land" and to brainstorm further possibilities.
*   She will use these initial reconciliations as a basis for further contacts. Hopefully her early contacts will also act as informal messengers to others within her family and among her former friends.

This is far as she wants to go for the moment in the planning process. She will to develop further plans only after a couple of successful contacts.

1. Before doing your own plan, get together with a learning partner and critique the plan that Emma has developed. What are its strong points? What are its deficiencies and how might they be remedied? What might you add to the plan?

The first goal you want to accomplish.

_____

_____

On a separate sheet of paper outline the major steps of your plan.

The second goal you want to accomplish.

_____

_____

On a separate sheet of paper outline the major steps of your plan.

EXERCISE 12-9: SHARING AND SHAPING PLANS

In this exercise you are asked to share and improve the plan you have developed, as well as help a learning partner do the same.

1. Share your written plans with a learning partner. Each of you should read the other's before you get together.
2. In a face-to-face session, explore and provide feedback to each other on the quality of each plan. How well shaped is it? To what degree is it simple without being simplistic? How concrete is it? What kind of compromise does it reach between too much and too little detail? How good is the fit between the plan and the person? Use empathic highlights, probes, and challenge to help the other shape his or her plan.
3. In the light of the feedback and dialogue, indicate what changes you would make in the plan.

The changes I would make in my first plan:

_____

_____

_____

_____

The changes I would make in my second plan:

_____

_____

_____

_____

168

EXERCISE 12-10: FORMULATING PLANS FOR THE MAJOR STEPS OF A COMPLEX GOAL

If a goal is complex-- for example, changing careers or doing something about a deteriorating marriage-- the plan to achieve it will often have a number of major steps. In this case, a divide-and-conquer strategy is useful. That is, the complex goal—say, the improvement of the marriage—can be divided up into a number of sub-goals. For instance, a more equitable division of household chores might be such a sub-goal. It is one goal in the total "package" of goals that will constitute the improved marriage. In this exercise you are asked to spell out the action steps for two of the sub-goals leading up to the accomplishment of some complex goal of your own. Consider this example.

Example: Lynette, a clinical psychology student in a counselor training program, has discovered that she comes across as quite manipulative, both to her instructors and to her classmates. She believes that she has developed the style as a response to the less than enthusiastic reception she received whenever she encroached on "male" territory, whether at home, school, or in the workplace. Her current style in working with others is to make the decisions herself while letting the other party think that he or she is having a say. Then, if she does not get her way, she moves to a more overtly domineering style. However, she realizes that this style is contrary to the values she wants to permeate her relationships, including, of course, her relationships with her future clients.

       Since this is the first time that she has received such feedback, one of the sub-goals she sets for herself is to get a very concrete understanding of her present style and to develop a better picture of a preferred style. This would be a major step in changing her overall manipulative style. Here are the steps in her plan to implement this style change:

- To elicit the cooperation of her instructor and the members of her training group. She wants to use the training experience as a lab for personal change.
- To spend three weeks interacting as she usually does. Meaning, for three weeks she will follow her instincts. She knows that she can't pull this off perfectly, because she is now observing herself more critically, as are the members of her group.
- To identify "live" instances when her domineering or manipulative style takes over. She can either catch herself or get immediate feedback from others when these instances take place. She will also keep track of instances of manipulative behavior in her everyday life.
- To identify "on the spot," briefly, how the interchange with another person or the group might have gone better. This will provide her with a practical inventory of possibilities for a change in style.
- At the end of three weeks, to draw up a portrait of the dysfunctional style and to pull together a portrait of the preferred style from the possibilities developed within the group.

Move on to the following tasks:

1. With a learning partner evaluate the steps she has laid out. If she were to ask you for your help, what feedback would you give her about her plan? What changes might you suggest?
2. Now do the same for two of the major steps of a plan you have developed to achieve some complex goal related to becoming a more effective helper.

A key goal of mine.

_____

_____

One major step in the plan to accomplish this goal.

_____

_____

_____

A brief description of the steps I would take to achieve this sub goal.

_____

_____

_____

_____

_____

_____

A second major step in the plan to accomplish this goal.

_____

_____

_____

A brief description of the steps I would take to achieve this second sub goal.

_____

_____

_____

_____

_____

170

3. Meet with a learning partner and swap finished exercises.
4. Get and give feedback on the viability of the plans. What would you change?

EXERCISE 12-11: DEVELOPING THE RESOURCES NEEDED TO IMPLEMENT YOUR PLANS

Plans can be venturesome, but they must also be realistic. Most plans call for resources of one kind or another. In this exercise you are asked to review some of the goals you have established for yourself in previous exercises and the plans you have been formulating to implement these goals with a view to asking yourself, "What kind of resources do I need to develop to implement these plans?" For instance, you may lack the kinds of skills needed to implement a program. If this is a case, the required skills constitute the resources you need.

Goal. Indicate a goal you would like to implement in order to manage some problem situation or develop some opportunity. Consider the following example. Mark is trying to manage his physical well-being better. He has headaches that disrupt his life. "Frequency of headaches reduced" is one of his goals, a major step toward getting into better physical shape. "The severity of headaches reduced" is another.
Plan. Outline a plan to achieve this goal. The plan may call for resources you may not have or may not have as fully as you would like. Consider Mark once more. Relaxing both physically and psychologically at times of stress, and especially when he feels the "aura" that indicates a headache is on its way, is one strategy for achieving his goal. Discovering and using the latest drugs to help him control his kind of headache is another part of his plan.
Resources. Indicate the resources you need to develop to implement the plan. For instance, Mark needs the skills associated with relaxing. Furthermore, since he allows himself to become the victim of stressful thoughts, he also needs some kind of thought-control skills. He does not possess either set of skills. Finally, he needs a doctor with whom he can discuss the kind of headaches he gets, and possible drugs available for helping control them.
Resources Plan. Summarize a plan that would enable you to develop some of the resources you need to move forward. In one program offered through the school, Mark learns the skills of systematic relaxation and skills related to controlling self-defeating thoughts. The Center for Student Services refers him to a doctor who specializes in headaches.

Problem situation #1: Your problem-managing or opportunity-developing goal.

_____

_____

_____

Skills or other resources you need to accomplish this goal.

_____

_____

_____

_____

Summarize a plan that can help you develop or get the resources you need.

_____

_____

_____

_____

_____

Problem situation #2. Your problem-managing or opportunity-developing goal.

_____

_____

_____

Skills or other resources you need to accomplish this goal.

_____

_____

_____

Summarize a plan that can help you develop or get the resources you need.

_____

_____

_____

_____

_____

Share your goals and your resource-development plans with a learning partner, and provide feedback to each other.

Chapter 13

THE ACTION ARROW: MAKING IT ALL HAPPEN

The Action Arrow of the helping model indicates the difference between planning and action. Stages I, II, III and their nine tasks all revolve around planning for change, not change itself. However, the need to incorporate action into planning, and planning into action, has been emphasized throughout the book. The "little actions" needed to get the change process moving right from the start have been noted and illustrated. We now take a more formal look at results-producing action. It is important to identify both the obstacles to action and ways to overcome these obstacles.

Stages I, II, and III are all planning stages. They are all just a lot of hot air if the client does not engage in *and* persist in the kind of activity that accomplishes goals. Some clients, once they have a clear idea of what to do to handle a problem situation – whether or not they have a formal plan – go ahead and do it. They need little or nothing in terms of further support and challenge from their helpers. They either find the resources they need within themselves or get support and challenge from the significant others in the social settings of their lives. At the other end of the spectrum are clients who choose goals and come up with strategies for implementing them, but who are, for whatever reason, stymied when it comes to action. Most clients fall between these two extremes. There are many reasons why clients fail to act on their own behalf:

- helpers who do not have an action mentality, who help clients discuss problems, but who do not help them act.
- client inertia, wherein clients fail to get started because change is hard work, and there are obstacles in the way of constructive change.
- client entropy, wherein the tendency of clients is to slow down and give up once they have started.

Often, a failure to see serious down-the-road obstacles to action contributes to both inertia and entropy. Chapter 13, which deals with these and other implementation issues, should be read before doing the exercises below.

Since the skilled-helper model is action-oriented, it is important for you to understand the place of "getting things done" in your own life. If you are to be a catalyst for problem-managing and opportunity-developing action on the part of clients, then reviewing your own track record in this regard is important. Unfortunately, helping often suffers from too much talking and not enough doing. Research shows that helpers are sometimes more interested in helping clients *develop* new insights than encouraging them to *act* on them. Inertia and procrastination plague most of us. The exercises in this section are designed to help you explore your own orientation toward action, so that you may become a more effective stimulus to action for your clients.

EXERCISE 13-1: EXPLORING YOUR OWN TENDENCY TO PROCRASTINATE

Procrastination, or putting things off, is part of the human condition. We all do it, at least sometimes. In fact, we can be become quite inventive in procrastination. Some procrastination is chronic – "I never seem to get around to doing papers for my courses until the last minute." But procrastination can also be incident or project specific – "I know I promised to go see my ailing aunt who loves me dearly, but I haven't gotten around to doing it." In this exercise you are asked to explore either chronic or incident-specific procrastination in your own life.

An example of chronic procrastination: Dahlia, a 52-year-old whose children, including one autistic son, are now grown, has returned to school in order to become a counselor. She has this to say about her tendency to procrastinate: "My husband is in business for himself. I take care of a lot of the routine correspondence for the business and our household. Often I let it pile up. The more it piles up the more I hate to face it. On occasion, an important business letter gets lost in the shuffle. This annoys my husband a great deal. Then, with a great deal of flurry, I do it all and for a while keep current. But then I slide back into my old ways. I also notice that when I let the mail pile up I waste a lot of time reading junk mail – catalogs of things I'm not going to buy, things I don't need. All of this is odd because now that I am in school time is at a premium and I need to become more efficient."

An example of incident-specific procrastination: Randolph, a 24-year-old counselor trainee, discusses a deferred reconciliation with a cousin who attended the same college, and with whom he had a falling out in his senior year. He says, "In high school, she had been the popular one. I was a late-bloomer, I suppose. I came to be very well-liked in college, while her social life was average. In senior year I found out that she had been telling lies about me behind my back. My success, I guess, galled her. I told her off and our relationship ended. It made family get-togethers difficult. We'd avoid each other, even though our families are close. So I found reasons to avoid family events. But I think we've both matured quite a bit over the last few years. I want to re-establish a relationship with her, but I keep finding reasons to put it off."

1. Describe an instance of chronic or project-specific procrastination in your own life.

_____

_____

_____

_____

_____

_____

2. What are the inhibitors? What keeps you from acting? What are the incentives for *not* acting?

_____

_____

_____

_____

_____

3. What available incentives would help you move to action? What can you do to mobilize them?

_____

_____

_____

_____

_____

4. Share your findings with a learning partner. What do you need to do to move beyond either your habit of procrastinating or the specific instance of procrastination you described?

EXERCISE 13-2: EXPLORING THE SELF-STARTER IN YOURSELF

While all of us have a tendency to put things off, we also have the possibility of become self-starters. Self-starters move to problem-managing and opportunity-developing action without being influenced, asked, or forced to do so. Skilled helpers are, ideally, self-starters. They have a sense of "agency" that enables them to help clients explore self-starting possibilities in themselves.

Example. Dahlia, whose chronic procrastination is described above, came up with the following list:

- "I never put off the things I like to do. For instance, I like the volunteer work I do at the hospital. No one has to put pressure on me to show up."
- "I make new friends easily. I don't wait for people to approach me. I take the initiative."
- "I plan family get-togethers and make them happen. Family solidarity is an important value for me. I do whatever I can to develop it."
- "I'm in no way a hypochondriac, but I do watch my health carefully. I get yearly check-ups, I see the dentist regularly, and if I experience symptoms that could mean something, I get myself checked out."

1. Describe four ways in which you are a self-starter.

_____

_____

_____

_____

_____

_____

175

2. Describe two ways in which you would like to become a self-starter.

_____

_____

_____

_____

_____

3. Share what you have learned about yourself as a self-starter with a learning partner. If you have trouble finding ways in which you are a self-starter or ways to become one, explore the implications of this with the training group.

EXERCISE 13-3: LEARNING FROM FAILURES

As suggested in the text, inertia and entropy affect all of us in our attempts to manage problem situations and develop unexplored opportunities. There is probably no human being who has not failed to carry through on some self-change project. This exercise assumes that we can learn from our failures.

Example: Miguel kept saying that he wanted to leave his father's business and strike out on his own, especially since he and his father had heated arguments over how the business should be run. He earned an MBA in night school, and talked about becoming a consultant to small family-run businesses. A medium-sized consulting firm offered him a job. He accepted on the condition that he could finish up some work in the family business. But he always found "one more" project in the family business that needed his attention. All of this came out as part of his story, even though his main concern was the fact that his woman friend of five years had given him an ultimatum: marriage or forget about the relationship.

Finally, with the help of a counselor, Miguel makes two decisions: to take the job with the consulting firm, and to break off the relationship with his woman friend, because he is still not seriously entertaining marriage as an immediate possibility. However, in the ensuing year Miguel never gets around to taking the new job. He keeps finding tasks to do in his father's company and keeps up his running battle with his father. Obviously both of them, father and son, were getting something out of this in some way. As to his relationship with his woman friend, the two of them broke it off four different times during that year, until she finally left him and got involved with another man.

Some of Miguel's learnings. Since Miguel and his counselor were not getting anywhere, they decided to break off their relationship for a while. However, when Miguel's woman friend definitively broke off *their* relationship, Miguel was in such pain that he asked to see the counselor again. The first thing the counselor did was to ask Miguel what he had learned from all that had happened, on the assumption that these learnings could form the basis of further efforts. Here are some of his learnings:

176

- "I hate making decisions that tie me down."
- "I pass myself off as an adventuresome, action-oriented person, but at root I prefer the status quo."
- "I am very ambiguous about facing the developmental tasks of an adult my age. I have liked living the life of a 17-year-old at age 30."
- "I enjoy the stimulation of the counseling sessions. I enjoy reviewing my life with another person and developing insights, but this process involves no real commitment to action on my part."
- "Not taking charge of my life and acting on goals has led to the pain I am now experiencing. In putting off the little painful actions that would have served the process of gradual growth, I have ended up in big pain. And I picture myself repeating the same pattern in the future."

Notice that Miguel's learnings are about himself, his problem situations, and his way of participating in the helping process. Now turn to your own life.

1. Recall some significant self-change project over the last few years that you abandoned in one way or another.
2. Picture as clearly as possible the forces at work that led to the project ending in failure, if not with a bang, then with a whimper.
3. In reviewing your failed efforts, jot down what you learned about yourself and the process of change.
4. Share your learnings with a learning partner. Help each other: (a) discover further lessons in the review of the failed project; and (b) discover what could have been done to keep the project going.

The self-change project that failed.

_____

_____

_____

_____

_____

Principal reasons for failure.

_____

_____

_____

_____

_____

_____

What you learned about yourself as an agent of change in your own life.

_____

_____

_____

_____

_____

_____

What could have been done to keep the project going?

_____

_____

_____

_____

_____

_____

EXERCISE 13-4: IDENTIFYING AND COPING WITH OBSTACLES TO ACTION

As suggested in an earlier exercise, "forewarned is forearmed" in the implementation of any plan. Identifying possible obstacles to a constructive change project is wisdom, not weakness.

1. Picture yourself trying to implement some action strategy or plan in order to accomplish a problem-managing goal. As in the example below, jot down what you actually see happening.
2. As you tell the story, describe the pitfalls or snags you see yourself encountering along the way. Some pitfalls involve inertia, or, not starting some step of your plan; others involve entropy, or, allowing the plan to run into a brick wall or fall apart over time.
3. Design some strategy to handle any significant snag or pitfall you identify.

Example: Chester has a supervisor at work who, he feels, does not like him. He says that she gives him the worst jobs, asks him to put in overtime when he would rather go home, and talks to him in demeaning ways. In the problem exploration phase of counseling, he discovered that he probably reinforces her behavior by buckling under, by giving signs that he feels hurt but helpless, and by failing to challenge her in any direct way. He feels so miserable at work that he wants to do something about it. One option is to move to a different department, but to do so he must have the recommendation of his immediate supervisor. Another possibility is to quit and get a job elsewhere, but, since he likes the company, that would be a drastic option. A third possibility is to deal with his supervisor more directly. He sets goals related to this third option.

178

One major step in working out this overall problem situation is to seek out an interview with his supervisor and tell her, in a strong but nonpunitive way, his side of the story and how he feels about it. Whatever the outcome, his version of the story would be on record. The counselor asks him to imagine himself doing all of this. What snags does he run into? Here are some of them:

- "I see myself about to ask her for an appointment. I see myself hesitating to do so because she might answer me in a sarcastic way. Also, others are usually around and she might embarrass me and they will want to know what's going on, why I want to see her, and all that. I tell myself that I had better wait for a better time to ask."
- "I see myself sitting in her office. Instead of being firm and straightforward, I'm tongue-tied and apologetic. I forget some of the key points I want to make. I let her brush off some of my complaints and in general let her control the interaction."

How can he prepare himself to handle the obstacles or snags he sees in his first statement? Then, what could he do about making sure he gets an appointment?

_____

_____

_____

_____

_____

How can he prepare himself to handle the pitfalls mentioned in his second statement? What could he do to deliver his messages in a decent but forceful way?

_____

_____

_____

_____

_____

Personal Situation # 1. Consider some plan or part of a plan you want to implement. In your mind's eye, see yourself moving through the steps of the plan. What obstacles or snags do you encounter? Jot them down.

_____

_____

_____

_____

_____

_____

Indicate how you might prepare yourself to handle a significant obstacle or pitfall and what you might do in the situation itself to handle it.

_____

_____

_____

_____

_____

_____

Personal Situation # 2. Consider some plan or part of a plan you want to implement. In your mind's eye, see yourself moving through the steps of the plan. What obstacles or snags do you encounter? Jot them down.

_____

_____

_____

_____

_____

_____

Indicate how you might prepare yourself to handle a significant obstacle or pitfall and what you might do in the situation itself to handle it.

_____

_____

_____

_____

_____

_____

EXERCISE 13-5: IDENTIFYING FACILITATING FORCES

In this exercise, you are asked to identify forces "in the field," or, out there in clients' day-to-day lives that might help them implement strategies and plans, together with forces that might hinder them. The former are called "facilitating forces" and the latter "restraining forces."

Example: Ira, as we have seen earlier, wants to stop smoking. He has also expanded his goal from merely "stopping" to "staying stopped." He has formulated a step-by-step plan for doing so. Before taking the first step, he uses force-field analysis to identify facilitating and restraining forces in his everyday life.

Some of the facilitating forces identified by Ira:

- my own pride.
- the satisfaction of knowing I'm keeping a promise I've made to myself.
- the excitement of a new program, the very "newness" of it.
- the support and encouragement of my wife and my children.
- the support of two close friends who are also quitting.
- the good feeling of having that "gunk" out of my system.
- the money saved and put aside for more reasonable pleasures.
- the ability to jog without feeling like I'm going to die.
- becoming more aware of the spiritual dimensions of life.

Now carry out the following tasks:

1. Review a goal or sub goal and the plan you have formulated to accomplish it.
2. Picture yourself "in the field" actually trying to implement the steps of the plan.
3. Identify the principal forces that are helping you reach your goal or sub-goal.

Spell out a goal or sub-goal you want to accomplish and they key steps you are taking to implement it.

_____

_____

_____

_____

Picture yourself in the process of implementing the plan formulated to achieve the goal. List the facilitating forces could help you to carry out the plan

_____

_____

_____

_____

_____

Which facilitating forces are most useful? What can you do to strengthen and use them? For instance, Ira focuses on three related facilitating forces: his pride, the satisfaction keeping commitments he makes, and feeling better by getting the "gunk" out of is system.

- Since he is a religious man, every morning he prayerfully recommits himself to his promise to quit.
- Every evening he congratulates himself for keeping his promises despite "flirtations with temptation."
- He periodically prides himself on becoming less dependent on chewing the gum that satisfies his craving for nicotine.
- He slowly and prudently increases his jogging time and "glories in the fresh air" in his lungs.

Now do the same for your action program.

_____

_____

_____

_____

_____

Finally, through dialogue share your findings with a learning partner.

EXERCISE 13-6: DEALING WITH RESTRAINING FORCES

Now turn your attention to whatever restraining forces there might be "in the field." When these are weak, they can be brushed aside. However, when they are powerful, they must be dealt with. We return to the case of Ira.

Some of the restraining forces identified by Ira:

- the craving to smoke that I take with me everywhere.
- seeing other people smoke.
- danger times: when I get nervous, after meals, when I feel depressed and discouraged, when I sit and read the paper, when I have a cup of coffee, at night watching television.
- being offered cigarettes by friends.
- when the novelty of the program wears off (and that could be fairly soon).
- increased appetite for food and the possibility of putting on weight.
- my tendency to rationalize and offer great excuses for my failures.
- the fact that I've tried to stop smoking several times before and have never succeeded.

Now do the following tasks:

1. Review a goal or sub-goal you identified in the previous exercise.
2. Once more, picture yourself "in the field" actually trying to implement the steps of the plan.
3. Identify the principal forces that are hindering you from reaching your goal or sub-goal.

List the restraining forces that might keep you from carrying out the plan.

_____

_____

_____

_____

_____

Which restraining forces are most powerful? What can you do to eliminate or minimize them? For example, Klaus is an alcoholic who wants to stop drinking. He joins Alcoholics Anonymous. During a meeting, he is given the names and telephone numbers of two people whom he is told he may call at any time, day or night, if he feels he needs help. He sees this as a critical facilitating force – just knowing that help is around the corner when he needs it. However, Klaus sees being able to get help anytime as a kind of dependency. He hates dependency, since he sees that its roots go deep inside him. This is a major restraining force. How does he handle it?

- First, he talks out the negative feelings he has about being dependent in this way with a counselor. The first step in managing this restraining force is to name it.
- In talking, he soon realizes that calling others is a temporary form of dependency that is instrumental in achieving an important goal: developing a pattern of sobriety. He sees calling as a safety value, not a continuation of the dependency streak within him.
- Next he calls the telephone numbers a couple of times when he is not in trouble just to get the feel of doing so. He likes letting others know that he is in good shape.
- He puts the numbers in his wallet, he memorizes them, and he puts them on a piece a paper and carries them in a medical bracelet that tells people who might find him drunk that he is an alcoholic trying to overcome his problem. He eliminates one excuse for not using the numbers.
- Finally, he calls the telephone numbers a couple of times when the craving for alcohol is high and his spirits are low. He gets practice in using this temporary resource.

Now do the same for your change program.

_____

_____

_____

_____

_____

_____

_____

_____

Finally, share your findings with a learning partner. Use empathy, probing, and challenge to help each other learn from the exercise.

EXERCISE 13-7: USING SUPPORTIVE AND CHALLENGING RELATIONSHIPS

Key people in the day-to-day lives of clients can play an important part in helping them stay on track as they move toward their goals. If clients are "out of community," then a parallel part of the helping process should be to help them develop supportive human resources in their everyday life. In this exercise you are asked to look at strategies and plans from the viewpoint of these human resources. People can provide both support and challenge.

Example: Enid, a 40-year-old single woman, is trying to decide what she wants to do about a troubled relationship with a man. She knows that she no longer wants to tolerate the psychological abuse she has been getting from him, but she also fears the vacuum she will create by cutting the relationship off. She is, therefore, trying to develop some possibilities for a better relationship. She also realizes that ending the relationship might be the best option. Because of counseling, she has been much more assertive in the relationship. She now cuts off contact whenever he becomes abusive. That is, she is already engaging in a

184

series of "little actions" that help her better manage her life and discover further possibilities. Over the course of two years, the counselor helps Enid develop human resources for both support and challenge.

- She moves from one-to-one counseling to group counseling with occasional one-to-one sessions. Group members provide a great deal of both support and challenge.
- She begins attending church. In the church she attends a group something like Alcoholics Anonymous.
- Through one of the church groups she meets and develops a friendship with a 50-year-old woman who has "seen a lot of life." She challenges Enid whenever she begins feeling sorry for herself. She also introduces Enid to the world of art.
- Enid does some volunteer work at an AIDS center. The work challenges her and there is a great deal of camaraderie among the volunteers. It also takes her mind off herself.

Since one of Enid's problems is that she is "out of community," these human resources constitute part of the solution. All these contacts give her multiple opportunities to get back into community and both give and get support and challenge. Now move on to the following tasks:

1. Summarize some goal you are pursuing and the action plan you have developed to get you there.
2. Identify the human resources that are already part of that plan.
3. Indicate the ways in which people provide support for you as you implement your plan.
4. Indicate ways in which people challenge you to keep to your plan or even change it when appropriate.
5. What further support and challenge would help you stick to your program?
6. Indicate ways of tapping into or developing the people resources needed to provide that support and challenge.

Summarize your goal and action plan.

_____

_____

_____

_____

_____

_____

Identify the human resources that are already part of that plan.

_____

_____

_____

_____

Indicate the ways in which people provide *support* for you as you implement your plan.

_____

_____

_____

_____

_____

Indicate ways in which people *challenge* you to keep to your plan or even change it when appropriate.

_____

_____

_____

_____

What further support and challenge would help you stick to your program?

_____

_____

_____

_____

_____

_____

186

Indicate ways of tapping into or developing the people resources needed to provide that support and challenge.

_____

_____

_____

_____

_____

_____